A FIELD GUIDE TO THE **Reptiles of Thailand**

A FIELD GUIDE TO THE

Reptiles of Thailand

TANYA CHAN-ARD

JOHN W.K. PARR, AND

JARUJIN NABHITABHATA

Illustrated by

Kritchanan Petcha-um,

Anantapong Poolsawat, *and*

Worachej Prom-Nu

OXFORD
UNIVERSITY PRESS

BRT
The Thai Response to Biodiversity

OXFORD
UNIVERSITY PRESS

Oxford University Press is a department of the University of Oxford.
It furthers the University's objective of excellence in research, scholarship,
and education by publishing worldwide.

Oxford New York
Auckland Cape Town Dar es Salaam Hong Kong Karachi
Kuala Lumpur Madrid Melbourne Mexico City Nairobi
New Delhi Shanghai Taipei Toronto

With offices in
Argentina Austria Brazil Chile Czech Republic France Greece
Guatemala Hungary Italy Japan Poland Portugal Singapore
South Korea Switzerland Thailand Turkey Ukraine Vietnam

Published in the United States of America by
Oxford University Press
198 Madison Avenue, New York, NY 10016

Library of Congress Cataloging-in-Publication Data
Parr, John W. K.
A field guide to the reptiles of Thailand / John W.K. Parr, Tanya Chan-ard, Jarujin Nabhitabhata ;
illustrated by Kritchanan Petcha-um, Anantapong Poolsawat, Worachej Prom-Nu.
pages cm
Includes bibliographical references and index.
ISBN 978-0-19-973649-2 (hardcover: alk. paper)—ISBN 978-0-19-973650-8 (pbk: alk. paper)
1. Reptiles—Thailand—Identification. I. Title.
QL661.T5P37 2015.
597.909593—dc23

9 8 7 6 5 4 3 2 1
Printed in China
on acid-free paper

Contents

Foreword

The publication of this field guide to the reptiles of Thailand fills a much-needed and valuable niche in Thai herpetology. Although the voluminous *The Lizards of Thailand* produced by Edward H. Taylor in 1963 provided an extremely valuable contribution to our knowledge of Thai reptiles, its publication in the comparatively obscure University of Kansas Science Bulletin indicates it was never intended as a field guide in the modern sense. Indeed the data found within its covers are so comprehensive, it resembles an *Encyclopaedia Britannica* of Thai reptiles.

The value of the current easy-to-follow guide to the field identification of all 352 reptiles currently known in Thailand lies in its conciseness. Its main aim is to stimulate a broader range of readers, reaching out to students, general interest enthusiasts, and amateur naturalists in the identification of the reptiles. Extremely ably assisted by the artistic skills by three of the nation's emerging wildlife artists, and supported by field observations from some of the world's most renowned field herpetologists, this attractive guide will undoubtedly assist in promoting the conservation of Thai reptiles and their habitats.

The publication of this book is extremely timely, as the reptiles of Thailand have never before been so threatened. Despite having an extremely impressive network of protected areas—perhaps one of the best in Southeast Asia—many critical lowland and riverine habitats for reptiles could not get protected status because of human and agriculture pressures. The merciless trade in wildlife has targeted some species, sending populations to the verge of extinction.

While the English language edition of this book will be well-received by visiting overseas tourists and keen wildlife enthusiasts alike, the knowledge that a sister Thai-language version is planned is refreshing. The opportunity to stimulate and foster interest among Thai citizens is long overdue.

Dr. Harold K. Voris,
Curator Emeritus
Department of Zoology
Field Museum of Natural History
Chicago, Illinois

Acknowledgements

Foremost we thank the National Science Museum. For providing support and facilities towards the production of this field guide the Biodiversity Research and Training Programme (BRT) for providing financial support towards the preparation of the plates.

We also thank the following people who imparted their knowledge on particular taxonomic groups quite freely: Dr. Uwe Fritz, Dr. John Iverson, Dr. William McCord, Dr. Peter Pritchard, Dr. Bryan Stuart, and Peter Paul van Dyke for assistance with the Testudines; Dr. Aaron Bauer, from the Department of Biology, Villanova University, for comments on *Cyrtodactylus*; Associate Professor Indraneil Das, from the Universiti Malaysia Sarawak on *Cnemaspis*; Dr. Jakob Hallermann of the Zoologisches Museum, Hamburg University, for his comments on the Agamids, and Prof. Dr. Wolfgang Böhme and Alexander Koenig of the Zoologisches Forschungsmuseum (ZFMK) as well as his Vietnamese student Nguyen Quang Truong for their assistance with the Scincidae. We thank Dr. Gernot Vogel of the Society for Southeast Asian Herpetology, Heidelberg, Germany. Special thanks are extended to Michael Cota, for his comments about Varanids and for his translations of German scientific papers.

Thanks to Van Wallach of the Museum of Comparative Zoology, Harvard University for his comments on the blind snakes. We particularly thank Dr. Vladimir E. Kharin, of the Institute of Marine Biology of Far East Branch of Russian Academy of Sciences, Vladovostok, for his generous assistance with the sea snakes; as well as Dr. Arne Redsted Rasmussen of the Royal Danish Academy of Fine Arts, School of Conservation, Copenhagen and Harold K. Voris, the Curator of Zoology, at the Field Museum of Natural History, Chicago, for their further advice on sea snakes.

A number of experts are currently studying Asian pit vipers. Among these, we extend our grateful thanks to Patrick David, of the Département Systématique et Evolution, Taxonomie-collection Reptiles & Amphibiens, Muséum National d'Histoire Naturelle, Paris, France; Dr. Anita Malhotra and Professor Roger Thorpe of the School of Biological Sciences, University of Wales, Bangor.

We thank Dr. Thosaporn Wongratana and Dr. Kumthorn Thirakhupt of the Biology Department, Chulalongkorn University for their assistance and provision of scientific papers.

John Parr extends his grateful thanks to his wife, Mukda, for her help in translating miscellaneous documents, and for her encouragement throughout.

How to Use this Book

Nomenclature

Little attention has been given to the nomenclature of common names of reptiles in the region. Most Asian countries have evolved lists of common names over the last century, often in isolation, which have then been reported in scientific literature. Some common names are misleading. Other names suggest an association to a particular country, even though the range is markedly larger. This field guide has opted to continue the usage of common names based on Nabhitabhata et al. (2000) as far as possible, to maintain consistency, but has taken account of developments in herpetology in the region. Further consideration has been given to the taxonomic reference.

Identification of Reptiles

The plates are the most important aid to the identification of most reptiles. When a reptile is encountered, the plate illustrations should be studied first until the closest match is found. With practice, and by reading the text, the main family groups can be recognized readily, greatly reducing the search effort.

Some species, such as cobras, show marked variations, particularly in coloration. Thus, the species accounts, which state the key identification features, should be consulted to confirm identification. Furthermore, a few reptilian species have been described from old museum specimens in which the colours may have altered or faded, so the colouring of live reptiles may appear slightly different.

Size is important in distinguishing many reptiles. Some indication of relative size is provided in the plates, but it is best to refer to the measurements provided within each species account. **Measurements** are provided which are most relevant to the morphology, and hence the identification, within each taxonomic group. The carapace measurements are provided for all turtles. The snout-to-vent measurements and tail measurements are provided for each of the lizard species. The body-length measurements are provided for all snakes. For the crocodilians, the length from head-to-tail is provided. These measurements are provided at the top right-hand side of each species account. These measurements are given in centimetres or metres, according to which is most appropriate to the respective reptile's size.

Reference should be made to the **Range Maps** provided for each species, as they may help both to confirm identification and to possibly eliminate similar species. These maps show the distribution of each reptile as accurately as possible based on collected museum specimens or reliable sightings of live animals. The distribution of both terrestrial and marine reptiles are shown in red.

The Species Accounts are divided into distinct sections. First, the key identification features are stated. The **Behaviour** of each species, as presently known, is provided to assist with identification. The **Habitat** is also stated for each reptile. For the habitats of forest-dwelling reptiles, the forest classification referred to later in the **Introduction** has been used. The **Range** provides the known range of the species in Thailand. The **Regional Range** provides a list of countries where the species has been recorded; note the range within these countries has not been defined a more specifically (e.g., as "southern Myanmar"), as the regional ranges continue to be refined as additional data are available.

Finally, the **International Conservation Status** is provided at the end of each species account. This status indicates (i) whether the species is threatened and (ii) whether the species is listed under the Convention for International Trade in Endangered Species, more commonly referred to as CITES.

With regard to threatened reptilian species, the categories are those now used in Red Data Books and Red Lists which have been in place, with some modification, for almost 30 years. Since their introduction these categories have become widely recognized internationally, and they are now used in a whole range of publications and listings, produced by the International Union for Conservation of Nature IUCN as well as by numerous governmental and non-governmental organizations. The Red Data Book categories provide an easily and widely understood method for highlighting those species under greater risk for extinction, to focus attention on conservation measures designed to protect them.

The categories are as follows:

Extinct (EX)—A taxon is Extinct when there is no reasonable doubt that the last individual has died.

Extinct in the Wild (EW)—A taxon is Extinct in the Wild when it is known only to survive in cultivation, in captivity, or as a naturalized population (or populations) well outside the past range. A taxon is presumed extinct in the wild when exhaustive surveys in known and/or expected habitat, at appropriate times (diurnal, seasonal, annual), throughout its historic range have failed to record an individual. Surveys should be over a time frame appropriate to the taxon's life cycle and life form.

Critically Endangered (CR)—A taxon is Critically Endangered when it is facing an extremely high risk of extinction in the wild in the immediate future.

Endangered (EN)—A taxon is Endangered when it is not Critically Endangered but is facing a very high risk of extinction in the wild in the near future.

Vulnerable (VU)—A taxon is Vulnerable when it is not Critically Endangered or Endangered but is facing a high risk of extinction in the wild in the medium-term future.

Lower Risk (LR)—A taxon is Lower Risk when it has been evaluated and does not satisfy the criteria as Critically Endangered, Endangered, or Vulnerable. Taxa included in the Lower Risk category can be separated into three subcategories:

- **Conservation Dependent (cd).** Taxa which are the focus of a continuing taxon-specific or habitat-specific conservation programme targeted toward the taxon in question, the cessation of which would result in the taxon qualifying for one of the threatened categories above within a period of five years.
- **Near Threatened (nt).** Taxa which do not qualify for Conservation Dependent, but which are close to qualifying for Vulnerable.
- **Least Concern (lc).** Taxa which do not qualify for Conservation Dependent or Near Threatened.

Data Deficient (DD) A taxon is Data Deficient when there is inadequate information to make a direct or indirect assessment of its risk of extinction based on its distribution and/or population status. A taxon in this category may be well-studied, and its biology well-known, but appropriate data on abundance and/or distribution are lacking. Data Deficient is therefore not a category based upon the level of threat. Listing a taxon in this category indicates that more information is required and acknowledges that future research will show that threatened classification is appropriate. It is important to make positive use of whatever data are available. In many cases great care should be exercised in choosing between DD and threatened status. If the range of a taxon is suspected to be relatively circumscribed, or if a considerable period of time has elapsed since the last record of the taxon, threatened status may well be justified.

Not Evaluated (NE) A taxon is Not Evaluated when it is has not yet been assessed against the criteria.

CITES is an international agreement between governments. Its aim is to ensure that international trade in specimens of wild animals and plants does not threaten their survival.

The widespread information available about the endangered status of many prominent species, such as tigers and elephants, might make the need for such a convention seem obvious. But when the ideas for CITES were first formed in the 1960s, international discussion of the regulation of wildlife trade for conservation purposes was something new. In hindsight, the need for CITES is clear. Annually, the international wildlife trade is estimated to be worth billions of dollars and to include hundreds of millions of plant and animal specimens. The trade is diverse, ranging from live animals and plants to a vast array of wildlife products derived from them, including food products, exotic leather goods, wooden musical instruments, timber, tourist curios, and medicines. The levels of exploitation of some animal and plant species are high and the trade in them, together with other factors such as habitat loss, are capable of heavily depleting their populations and even bringing some species close to extinction. Many wildlife species in trade are not endangered, but the existence of an agreement to ensure the sustainability of the trade is important in order to safeguard these resources for the future.

Because the trade in wild animals and plants crosses national borders, the effort to regulate it requires international cooperation. CITES was conceived in the spirit of such cooperation. Today, it accords varying degrees of protection to more than 30,000 species of animals and plants, whether they are traded as live specimens, fur coats, or dried herbs.

CITES was drafted as a result of a resolution adopted in 1963 at a meeting of members of IUCN (The World Conservation Union). The text of CITES was finally agreed at a meeting of representatives of 80 countries in Washington, DC, United States of America on March 3, 1973; on July 1, 1975, CITES entered in force.

CITES is an international agreement to which States (countries) adhere voluntarily. States that have agreed to be bound by the Convention (i.e.,"joined" CITES) are known as Parties. Although CITES is legally binding on the Parties—in other words, they must implement the Convention—it does not take the place of national laws. Rather, CITES provides a framework to be respected by each Party, which has to adopt its own domestic legislation to make sure that CITES is implemented at the national level.

Not one species protected by CITES has become extinct as a result of trade since the Convention entered into force and, for many years, CITES has been among the largest conservation agreements in existence, with more than 150 Parties.

How CITES Works

CITES works by subjecting international trade in specimens of selected species to certain controls. These controls require that all import, export, reexport and introduction from the sea of species covered by the Convention has to be authorized through a licensing system. ("Re-export" means export of a specimen that was imported.)

The species covered by CITES are listed in three Appendices according to the degree of protection they need.

Appendix I includes species threatened with extinction. Trade in specimens of these species is permitted only in exceptional circumstances.
Appendix II includes species not necessarily threatened with extinction, but in which trade must be controlled in order to avoid utilization incompatible with their survival.
Appendix III contains species that are protected in at least one country, which has asked other CITES Parties for assistance in controlling the trade.

Glossary

Acrodont dentition: Teeth are fused on to the dorsal surface of jaw bone

Anal plate: Terminal ventral scale or scute

Anapsid: Refers to a reptilian skull in which there are no temporal openings, known only in turtles today.

Anterior: Front part of an animal

Arboreal: Living in trees

Aquatic: Living in water

Axilla: The cavity beneath or behind the insertion of the anterior appendage, or armpit

Basal: At or near the base

Bridge: Part of shell in turtles which connects the carapace and plastron

Carapace: The upper shell of the turtle

Carnivorous: Diet consisting primarily of meat

Chevron: An inverted V-shaped marking

Communal: Lives in groups

Concave: bent inward, rounded

Convex: bent outward, rounded

Crepuscular: Active at twilight (dusk) or dawn

Dermal: Of the skin

Dewlap: A loose flap of skin under the throat

Diastema: A space or gap between two teeth

Diurnal: Active during the day

Dorsal: Relating to the back of the body

Forelegs: Front legs

Fossorial: Adapted for digging

Frontoparietal: Plate or plates lying on dorsum of head between the frontal and the parietals in lizards.

Groin: The angle formed by the anterior margin of the hind limb and the body; the sight depression or cavity at the insertion of the hind limb

Gular: Pertaining to or located on the throat

Herbivorous: Diet consisting primarily of plants

Infralabial: Lower labial

Infra-marginals: A lamina lying between the marginals of the carapace and the lateral margin of the normal plastral laminae

Intergular: A single or one of a pair of laminae lying between the gular laminae of turtles
Lateral On the side of an animal's body
Laterodorsal: The outer part of the back, usually a dark stripe
Loreal: A plate lying on side of head between nasals and preoculars
Marginals: A lamina in the outer series on the carapace of the turtle, visible both from above and below, and characterized by a sharp angle marking the edge of the bony part of the carapace
Nocturnal Active during the night
Nostril: One of the two channels of the nose
Nuchal: Relating to the neck
Occipital: Head scales posterior to and bordering on the parietals and interparietal (if present), particularly where the scales are distinct from other scales in vicinity
Omnivorous Feeds on both meat and vegetable matter
Osteoderm: A bony deposit in the form of a plate or a scale found in the dermal layers of the skin
Palatine: One of a pair of dermal bones in the roof of the mouth, lying just behind the vomers, and usually separated by a vomer-pterygoid contact
Patagium: Fold of skin on flanks supported by ribs
Pelagic: The entire volume of ocean wateer; subdivided into the neritic zone (relatively shallow waters overlying continental shelves) and the oceanic zone (water over ocean basin)
Piscivorous: Feeds on fish
Plastral: A general or group term for any one of the series of laminae covering the plastron in turtles
Plastron: Lower part of a turtle shell
Pleurals: Any one of the bony plates covering the ribs in the turtle carapace
Postorbital arch: A bar of bone extending from the roof of the skull to the upper lip line in those forms which have deeply excavated temporal openings, such as lizard
Prefrontal: A cranial dermal covering bone, found on the anterior margin of the orbit, and usually bounded dorsally by the frontal and laterally by the lacrymal
Premaxillary: Either of the two dermal bones lying at the anterior end of the skull, meeting on the midline and forming the anterior angle of the upper jaw
Proboscis: An elongate and flexible extension of the snout
Procoelian: A single vertebra
Pterygoid: Either of a pair of dermal bones forming a major part of the palate in many species
Reticulated: Forming a net-like pattern or reticulum
Rostral: The scale at the tip of the snout
Scansorial: Both arboreal and terrestrial
Serrated: Having a saw-like or jagged edge

Shields: A large scale in snakes and turtles
Striations: Having lines along the body
Supracaudal: Postcentral lamina on the turtle carapace
Superciliary: Small scale bordering orbit
Supralabial: Upper lip scale
Supranasal: Scale or scales lying immediately above the nasal and lateral to the internasal
Temporal: Scale or scales behind the postoculars, below the parietal, and and above the upper labials
Terrestrial: Living on the ground
Tympanum: The membrane covering the external opening of the middle ear chamber or vestibule
Underparts: Lower parts of an animal's body
Underside: Lower part of an animal's body
Upperside: Upper part of an animal's body
Ventral: Relating to the underside
Vertebral: Any one of the row of scales lying on the mid-dorsal line of the body

KEY TO THE MAJOR GROUPS OF THAI REPTILES

1	Skin moist and naked, not covered with cornified scales, scutes, or dermal plates	**Amphibians**
•	Skin dry, cornified and divided into scales or scutes, or formed into dermal plates	**2 (Reptiles)**
2	Body enclosed in a bony or a soft, fleshy shell	**Turtles and tortoises**
•	Body without a shell	**3**
3	Body covered with large dermal plates; tail compressed with two dermal ridges formed by dermal plates	**Crocodiles**
•	Body covered with cornified scales; tail cylindrical, flattened, or compressed, but without two dorsal ridges formed by dermal plates	**4**
4	Body long and slender; without legs or with only vestiges of hind legs (represented by pair of spur-like structures next to vent); no eyelids or ear openings; tail shorter than body	**Snakes**
•	Body short and stout or long and slender; with four legs or reduced to two or none (if none, then eyelids and ear openings present; body with deep, flexible groove on sides); tail equal to or longer than snout-vent length (unless regenerated)	**Lizards**

Modified from the key prepared by Zhao and Adler (1993).

Distinguishing features for the Identification of Thai Reptiles

In many Thai reptiles, the body size, shape, colour, and patterning are diagnostic. In the lizards and snakes, the scale patterns and scale counts on the head and body are often important for identification.

Turtles, Terrapins, and Tortoises:—the carapace and plastron

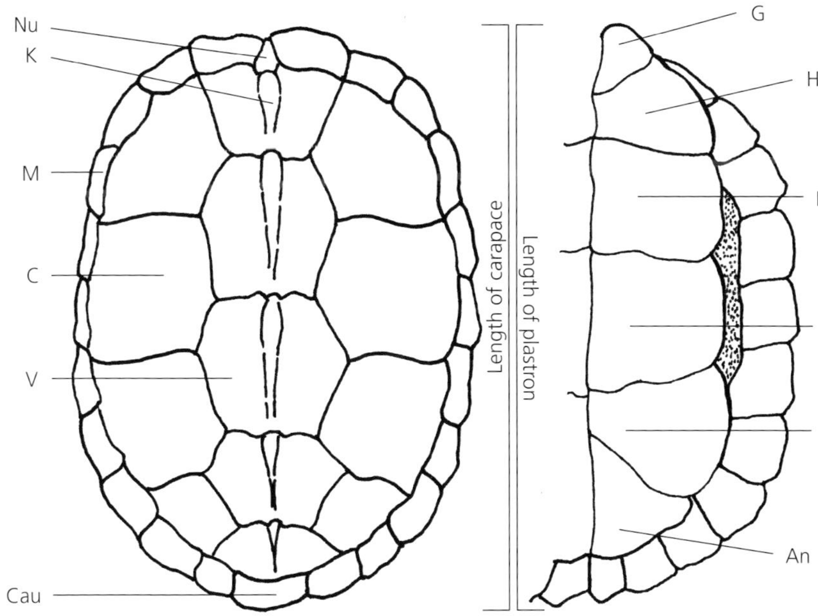

Carapace | **Plastron**

Nu	Nuchal
K	Keel
M	Marginal
C	Costal
V	Vertebral
Cau	Caudal

G	Gular
H	Humeral
P	Pectoral
Ab	Abdominal
F	Femoral
An	Anal

The scales (scutes) on the upper shell (carapace) and the lower shell (plastron) of hard-shelled turtles, terrapins and tortoises are arranged in a specific pattern which may assist identification.

Lizard body: indicating measurement points

All lizards are measured from the snout to the vent (SVL), rather than the total length. This is because lizards can lose and then regenerate their tails, rendering the use of total body length as an identification feature impractical.

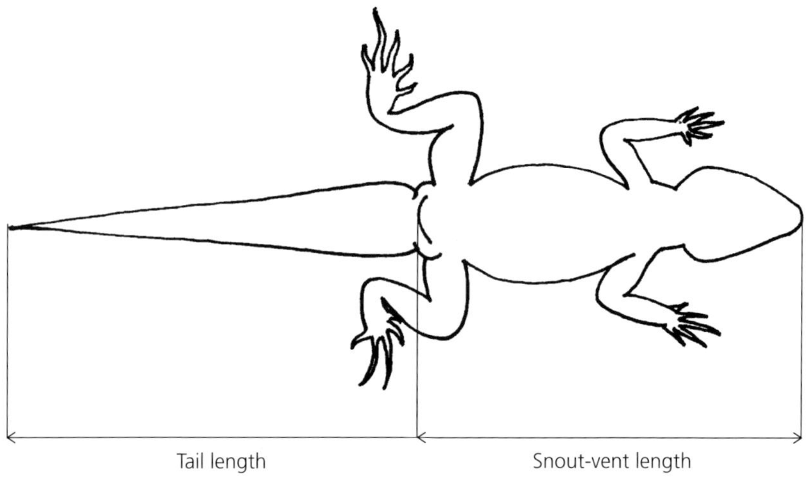

Skinks: the head showing the head scales

Scale position and size are very useful identification features for lizards. In all lizards, the position and name of these scales are more or less similar.

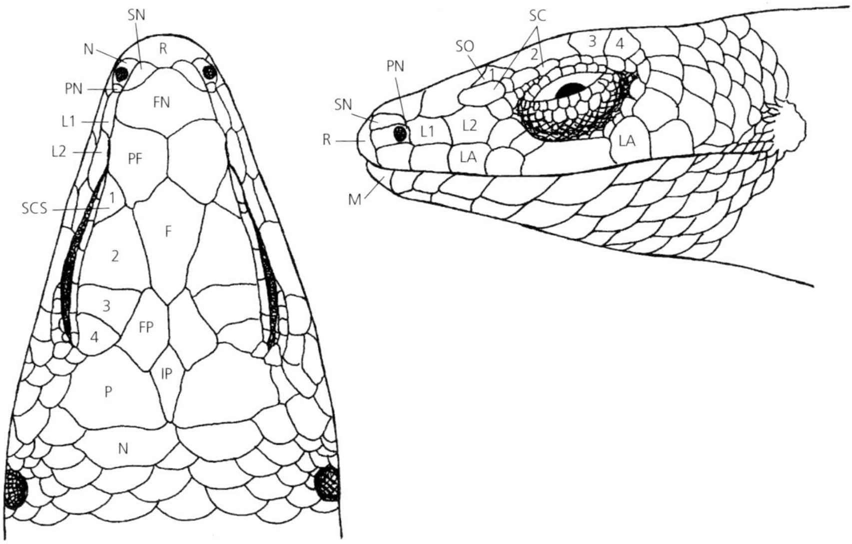

Head of *Mabuya multifasciata*,
to explain the nomenclature of the head-shields.

F. Frontal.
FN. Fronto-nasal.
FP. Fronto-parietal.
IP. Interparietal.
L. Loreals.
LA. Upper labials.
M. Mental.
N. Nasal.
N. Nuchal.
P. Parietal.
PF. Prefrontal.
PN. Postnasal.
R. Rostral.
SC. Supraciliaries.
SN. Supranasal.
SO, SCS. Supraoculars.

Gekkos:—the anal region

In geckos, the scale position and size in the anal region are very useful identification features.

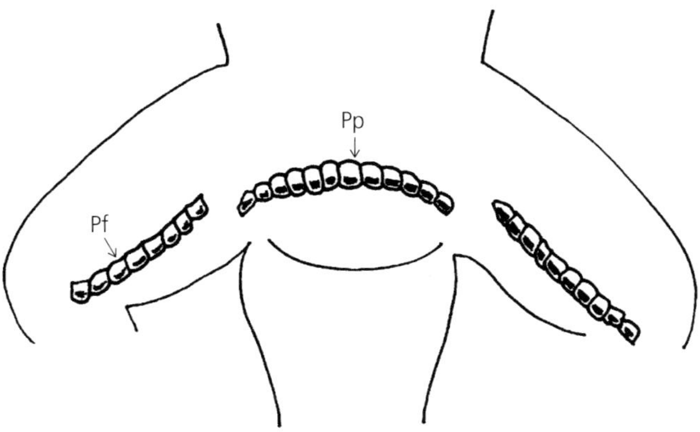

Pf – femoral pores
Pp – preanal pores

Snakes:—the head showing head scales

The scales on the heads of snakes can be identified by their shape and position.

Head scales of snake.

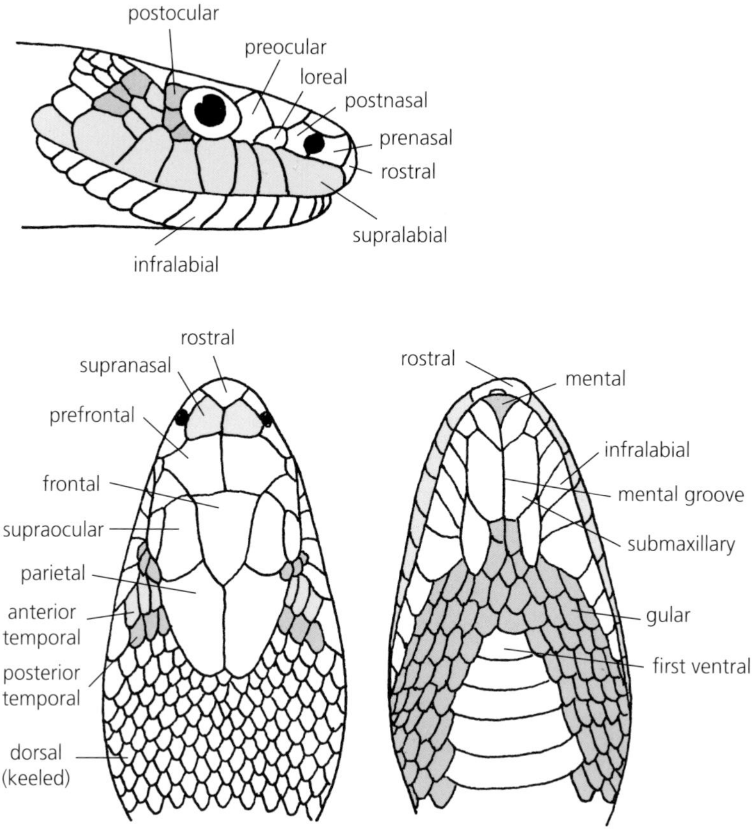

Snakes:—the method of counting the number of scale rows

Counts of the number of scale rows on a snake's body are useful for identifying species. The body scale count is made halfway between the head and the cloaca, where the number of rows is the highest. The ventral scale rows are not included in the count.

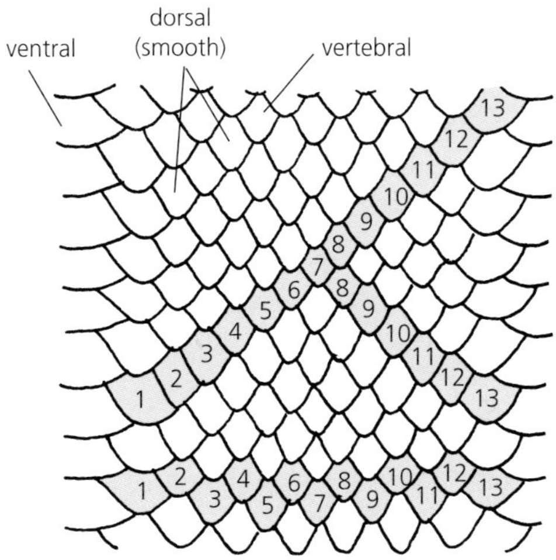

Methods of counting number of scale

Snakes:—the scales of the anal region

The sub-caudal scale count is the number of scales, or pair of scales, under the tail. The count is made from the first scale or pair of scales below the anal scale, to the scale just in front of the terminal scale at the tip of the tail.

Scales of the anal region.

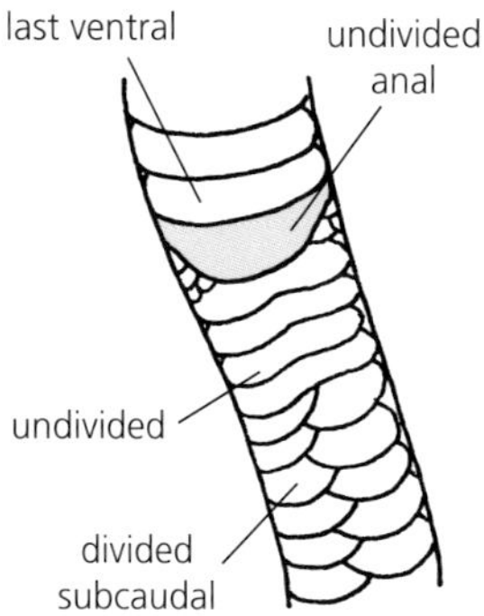

A FIELD GUIDE TO THE **Reptiles of Thailand**

Introduction

Interest in herpetology in Thailand is extremely limited, with perhaps fewer than 20 practicing herpetologists in the country. Indeed, the authors of a recent scientific paper describing a new record of a keelback from northwestern Thailand, stated that the herpetofauna of Thailand, "remains very poorly known." Part of this is a reflection of the history of the subject which has developed in the last two centuries. But it is also predominantly the result of the lack of user-friendly publications available to both amateur naturalists and professional wildlife researchers. Additionally, the rich diversity of similar-looking species makes identification no simple matter, putting off large numbers of potential practitioners.

Field guides are essential tools for stimulating interest in the different taxonomic groups. The identification and naming of the animals which we encounter represent a crucial component in generating broad interest among the general public in wildlife conservation. The lack of these field guides, particularly in the native Thai language, is a serious impediment to conservation efforts. However, the production of field guides can contribute vastly toward stimulating deep-rooted interest in the life forms which constitute our natural heritage. This guide includes species accounts for all of the 352 reptile species currently with confirmed records in Thailand, as of December 31, 2010.

History of Herpetology in Thailand

Our knowledge of the reptiles occurring in Thailand, including their distribution and ecology, has been extremely patchy. Initially, it was compiled from anecdotal contributions from a number of expatriates present in the Asian region from the 1700s through to the turn of the 20th century. During King Narai's reign (1656 to 1688), diplomatic missions were sent and received. The French court sent a mission under Chevalier de Chaumont to Ayutthaya ostensibly to convert King Narai to Catholicism. Within this mission, a Jesuit named Guy Tachard (1648–1712) first came to Siam (the name for Thailand before June 1939), with the Chaumont-Choisy embassy of 1685 as one of six Jesuits destined for missionary work in China. Tachard made some of the first observations of reptiles in Thailand.

Almost a century later, Sir John Bowring (1792–1872) was an English political economist, traveler, a writer on miscellaneous issues, and polyglot. In 1849, Bowring was appointed British consul at Canton (today's

Guangzhou), and superintendent of trade in China, a post he held for 4 years. On April 13, 1854 Bowring was sent to Hong Kong as governor. In 1855 he visited Siam, and negotiated with King Mongkut a treaty of commerce, now commonly referred as the Bowring treaty. He made further observations on the reptiles in Thailand. Bowring retired in March 1859 and received a pension for his duties.

In the mid-19th century, our knowledge of Thai herpetology expanded through a combination of collectors working in the field supported by taxonomists working out of museums in western Europe. Albert Carl Ludwig Gotthilf Günther (1830–1914) was a German-born British zoologist, who went to the British Museum in 1856, where he worked on ichthyology. Following the death of John Edward Gray in 1875 Günther was made director of the Zoology Department of the Natural History Museum, where he remained until 1895. In addition to fishes, Günther worked on the reptiles and amphibians in the museum's collection. Henri Mouhot (1826-1861) was an artist, explorer, and naturalist. He traveled extensively in Thailand, Cambodia, and Lao People's Democratic Republic (PDR) between 1858 and 1861, finally succumbing to malaria near Luang Prabang. The reptiles Mouhot collected were described by Günther and are in the British Natural History Museum.

George Albert Boulenger (1858–1937) has been called the leading taxonomic herpetologist of his era, and was probably the greatest of the 19th-century contributors to Asian herpetology. Originally from Brussels, Boulenger was appointed Assistant Naturalist at the Musee Royal d'Histoire Naturelle de Belgique in 1880. In this capacity, he made frequent trips to Paris and London, and it was during his trips to London that he drew the attention of Günther, the keeper of Zoology at the British Museum. Gunther was very impressed with Boulenger, and hired him in 1881 as assistant in charge of the lower vertebrates, assigning him the task of preparing a new edition of the catalogues of amphibians and reptiles. This resulted in nine volumes (covering 8,469 species) being published between 1882–1896; it remains the most comprehensive review of the world's herpetofauna ever produced.

Boulenger's work on the reptiles and amphibians of the Malay Peninsula, *A Vertebrate Fauna of the Malay Peninsula from the Isthmus of Kra to Singapore Including Adjacent Islands: Reptilia and Batrachia* (1912), in conjunction with Malcolm A. Smith's supplement of 1931, established the basic taxonomic framework for the herpetology of the Malay Peninsula. The work still stands, almost a century later, as the most valuable detailed systematic review of our herpetofauna.

Frank Wall (1868-1950) was a British medical officer and leading student of snakes who lived in Sri Lanka and India. His father worked there and was responsible for initiating the study of natural history on the island. Wall studied medicine in London and joined the Indian Medical Service in 1893. Sent to India under the British Raj, Wall continued to work there until 1925 and researched many animals, especially snakes. He collected numerous snakes, many of which are now in the British Museum. Wall was a member of the Bombay Natural History Society and published more than 200 scientific articles, as well as the book *A Popular Treatise on the Common Indian Snakes*. He died in Bournemouth. Another notable expatriate in the region was Stanley Smyth Flower (1871–1946) who was an avid naturalist. On leaving college at the age of 18 he joined the Northumberland Fuseliers, where he served in India and the Straits Settlement for 6 years.

Thai herpetology owes much to two individuals in particular.

Malcolm Arthur Smith (1875–1958) was a herpetologist working in the Malay Peninsula. By profession he was a physician. Smith was the founder and first president of the British Herpetological Society, in 1947. He published *Monograph of the Sea-snakes* (1926); and the *Fauna of British India. 3 volumes on Reptiles and Amphibians* (1931–1943). In 1951, Smith published *The British Amphibians and Reptiles.* Along with George Albert Boulenger, Smith laid the basic framework for herpetology in the Malay Peninsula. With a strong interest in reptiles and amphibians since childhood, Smith took up medicine as a practical means of living that would allow him to indulge in his passion of herpetology. Initially working as a medical officer with the British Legation in Bangkok, and taking up additional duties to the Royal Court of Thailand, in 1925 Smith was given bench space at the British Museum of Natural History as a platform to pursue his primary interests.

Working from his own collections as well as others, Smith published profusely on the herpetofauna of a wide range of regions—initially from the Malay Peninsula, Borneo, Hainan Island, and Indochina, and later from India, Burma, Indo-Australia, Australia, Afghanistan, and Nepal as well. His most important work in general is acknowledged to be the three volumes on reptiles and amphibians in the *Fauna of British India* series, but, in the context of the amphibians of the Malay Peninsula, his work *The Reptilia and Amphibia of the Malay Peninsula*, a supplement to Boulenger's *A Vertebrate Fauna of the Malay Peninsula from the Isthmus of Kra to Singapore including Adjacent Islands: Reptilia and Batrachia* of 1912, stands together with the latter as the basic reference for Malayan herpetology.

Dr. Edward Harrison Taylor (1889-1978), a professor at Kansas University, devoted his research to the animals of the Philippines, especially herpetofauna. He conducted several extended visits to Thailand (from September 1957 until June 1958, from July 1959 until September 1960, and in 1961) exploring and collecting in the country or along the borders with neighboring countries. Several thousand specimens were made available to Taylor for study. From these specimens and a study of specimens kept in various collections in a number of European and America Museums he wrote two publications *The Lizards of Thailand* (1963) and *The Serpents of Thailand and Adjacent Waters* (1965), published in the *University of Kansas Science Bulletin*. Taylor built up his own personal research library over a period of 50 years which was one of the finest private herpetological research libraries in the world.

Currently, there are about 20 Thai nationals with a keen interest in herpetology supported by a network of more than 100 experts in particular fields of hepetological taxonomy. The overwhelming number of these experts have contributed to the advancement of our understanding of the rich diversity of turtles, lizards, snakes, and crocodilians in the country. Many of these experts have reviewed sections of this field guide.

Physiography

Thailand lies in the Indo-Chinese Peninsula of the Oriental (or Indo-Malayan) faunal region, covering a land area of approximately 513,115 km^2. It extends from 5°37' N latitude to 20°30' N latitude and between 97°20' and 105°39' E longitude (approximately 1500 km from north to south and 800 km from east to west).

Thailand may be divided into four principal geological regions. The first region comprises the mainly mountainous north and west with four large rivers flowing through it. These waterways in turn drain into the second region, the alluvial Central Plains, which extends to the Gulf of Thailand. The northeast (usually referred to as the Khorat Plateau) constitutes the third region, being an area of dry, poor soils derived from sands, clays, and salt deposits. This region drains into the Mekong River which separates Thailand from Lao PDR. The fourth region is the Peninsula, a region dominated by a mountainous backbone, which is itself an extension of the mountain spine which extends down the west of the country. Throughout this guide, the main land mass of the country (including the Peninsula north of Prachuap Khiri Khan) is referred to as continental Thailand.

Climate

Thailand has a tropical, monsoonal climate with a distinct wet season and a long, hot dry season. Most parts of the country receive more than four-fifths of their annual rainfall during the southwestern monsoon, from May to October. Some areas of the Peninsula, particularly on the eastern coast, and in the extreme southwestern provinces, receive some additional rainfall from the northeast monsoon during November to January. The least seasonal areas are in the southernmost provinces of Pattani, Yala, and Narathiwat which, although they receive rain in both monsoons, are actually wettest in November and December.

The driest, most seasonal areas lie in the very centre of the Khorat plateau, in the country's northeastern "bulge," and in some parts of the west, which lie in the rain shadow of the mountains along the Burmese border. The average annual rainfall of some of these areas is only about 1000 mm.

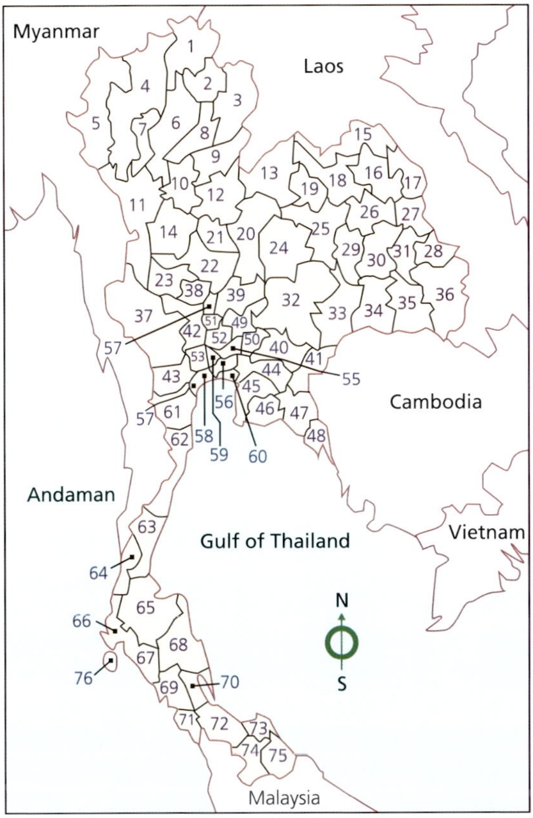

Provinces

1. Chiang Rai
2. Phayao
3. Nan
4. Chiang Mai
5. Mae Hong Son
6. Lampang
7. Lampun
8. Phrae
9. Uttaradit
10. Sukhothai
11. Tak
12. Phitsanulok
13. Loei
14. Kamphaeng Phet
15. Nong Khai
16. Sakhon Nakhon
17. Nakhon Phanom
18. Udon Thani
19. Nong Bua Lamphu
20. Petchabun
21. Phichit
22. Nakhon Sawan
23. Uthai Thani
24. Chaiyaphum
25. Khon Kaen
26. Kalasin
27. Mukdahan
28. Amnat Charoen
29. Maha Sarakham
30. Roi-Et
31. Yasothon
32. Nakhon Ratchasima
33. Buriram
34. Surin
35. Si Saket
36. Ubon Ratchathani
37. Kanchanaburi
38. Chai Nat
39. Lopburi
40. Prachinburi
41. Sa Kaeo
42. Suphanburi
43. Ratchaburi
44. Chachoengsao
45. Chonburi
46. Rayong
47. Chanthaburi
48. Trat
49. Saraburi
50. Nakhon Nayok
51. Ang Thong
52. Ayuthaya
53. Nakhon Pathom
54. Singburi
55. Pathum Thani
56. Bangkok
57. Samut Songkhram
58. Samut Sakhon
59. Nonthaburi
60. Samut Prakan
61. Phetchaburi
62. Prachuap Khiri Khan
63. Chumphon
64. Ranong
65. Surat Thani
66. Phang Nga
67. Krabi
68. Nakhon Si Thammarat
69. Trang
70. Phatthalung
71. Satun
72. Songkhla
73. Pattani
74. Yala
75. Narathiwat
76. Phuket

The wettest areas lie in the extreme southeast and on the west coast of the Peninsula, where the average rainfall is 3000 to 4000 mm per year.

The average annual temperature over most of the country lies in the range of 26-29°C, but there is considerable variation, with the more land-locked parts of the country displaying more variation than the coastal regions. In the hottest month, usually April, daytime temperatures in the lowlands may rise to more than 40°C. In the coolest months of December and January, ground frosts are frequently recorded on exposed ridges on the higher mountains of the north.

These marked regional variations in climate have a direct influence on the vegetation found in the different parts of the country. The vegetation, as habitats, provide the food sources of mammalian species.

Habitats

Terrestrial Forests

The majority of reptilian species are confined to the forests of Thailand. However, the distribution of forest types is complex, being primarily influenced not only by climate, but also topography and soil types. However local factors such as drainage, exposure, proximity to watercourses, and other factors also influence the distribution of tree species.

Two general types may be distinguished, namely *evergreen forest* and *deciduous forest*. Evergreen forests, in which the leaves remain green all year round, are found in climates with a dry season of less than three months, in areas with mean annual rainfall greater than 1500 mm; such forests may be found up to the highest peaks in the country. By comparison, deciduous forests, in which the trees shed their leaves as a response to water stress during the dry season, are found in climates with more than 3 months of drought and less than 1500 mm mean annual rainfall. They are generally found in the lowlands, but may ascend to 1200 m.

We use the following habitat classification in this book, which incorporates forest categories suggested by the Royal Forest Department (1991) in *Plants for our Future: Botanical Research and Conservation Needs in Thailand*.

Evergreen Forests

Malayan Mixed Dipterocarp Forest

This tall lowland forest occurs in climates lacking a regular dry season. It has a distinct discontinuous emergent stratum above the main canopy, mainly composed of dipterocarps. It is exceedingly rich in tree species, principally due to the presence of large genera containing series of coexisting species. Flowering in the canopy is supra-annual. This forest is confined to the extreme southern provinces.

Monsoonal Evergreen Forest

This tall lowland forest is unique to Indo-Burma, with a distinct discontinuous emergent stratum above the main canopy, in which dipterocarps are generally mixed with other species, especially in Bombacaceae, Sterculiaceae, and Leguminosae. This forest type is rich in tree genera and species, many of which are endemic. It occurs up to about 800 m in areas where annual rainfall exceeds 2000 mm and where the dry season is less than three months in duration. Flowering is more or less annual. There are two geographical types, differing in species composition and soils, and largely in geographical distribution.

Peninsular Monsoonal Evergreen Forest

This forest is found from Chumphon Province southward, usually on clay-rich soils derived from granite. Characteristic species of this forest type include *Parashorea stellata, Shorea farinosa, Dipterocarpus grandiflorus, D. kerrii,* and *Intsia palembanica.* Some emergents species are shortly deciduous.

Southeastern Monsoonal Evergreen Forest

This evergreen forest is found in extreme southeastern Thailand. Characteristic species include *Dipterocarpus costatus, Podocarpus wallichii, Shorea hypochra,* and *Swintonia pierrei.* The emergent canopy is almost entirely evergreen on yellow sandy leached soils derived from sandstone. A distinct facies occurs on podzols.

Dry Evergreen Forest

This forest occurs in seasonal climates with between 3 and 5 dry months and more than 1500 mm mean annual rainfall, and as gallery forest in the driest Thai climates, often with *Dipterocarpus alatus* dominant. It is similar to wet seasonal evergreen forest, but is substantially poorer in species, shorter in stature, and with more- or less-abundant deciduous species in genera such as *Lagerstroemia, Afzelia*, and including *Pterocarpus macrocarpus* and *Sindora siamensis*; which also occur in mixed deciduous forest. Characteristic emergent species of this zonal type include *Dipterocarpus turbinatus, Hopea odorata* and *Shorea henryana.* Fire may occasionally penetrate this forest, temporarily replacing the understorey with bamboo.

Montane Forests

These forests are of low to moderate stature, with a more or less even canopy structure, generally lacking emergents. They are moderately rich in tree species. Variable in composition but with Fagaceae, Lauraceae, and Theaceae dominant. Evergreen forests occur from 800 m—the diurnal cloud base—upward. Two altitudinal types may be recognized:

Lower Montane Forest

This forest type is found from 800 to 1800 m. It tends to be taller than Upper Montane forest and is richer in tree species and poorer in epiphytes. A distinct facies on dry ridges is prone to lightening strikes and is dominated by *Pinus kesiya*, an emergent, below which there is usually a grass field layer.

Upper Montane Forest

This forest is found above 1800 m, being confined to the highest peaks. It is distinguished by its shorter stature, poorer tree flora, and abundant epiphytes, including mosses and ferns. Many species are local or regional endemics.

Limestone Forest

Limestone forests are found throughout the country, forming karst topography. These karsts are devoid of soil and subject to extreme water stress. The forest is low in stature, except on basal screes, and with an even canopy. Deciduousness is variable, but high on the driest peaks and in the most seasonal climates. The calcium-rich conditions support a distinctive flora rich in endemic ground herbs, palms and some woody-leaved genera. Due to their precipitous topography, they generally remain well-forested.

Peat Swamp Forest

Peat swamp forests are found in everwet equatorial climates, where the absence of a dry season permits oligotrophic peat accumulation. It is a distinctive evergreen forest of moderate stature. Confined to the southern peninsula, these forests were once far more extensive, but have been all but entirely lost. Of note, an 80 km^2 remnant is found at Chalerm Pa Kiet Wildlife Sanctuary (or Pa Phru) in Narathiwat province.

Mangrove

A low stature, tropical evergreen forest with unique flora which show a number of specialized adaptations which enable them to survive inundation by brackish and saline waters. It is backed by a belt subject to highest seasonal tides but otherwise dry. It comprises very few species, and is low in endemics. They are found in sheltered inlets and estuaries, often in association with extensive sea-grass beds and intertidal mudflats.

Deciduous Forests

These forests are found in the lowlands, only occasionally up to 1200 m, in climates with more than three months of drought and less than 1500 mm mean annual rainfall. Humans, by means of fire, has extended these forests at the cost of the evergreen types, so that their original extent is impossible to ascertain.

Mixed Deciduous Forest

These forests are found on clay rich, relatively fertile soils throughout continental Thailand. They are forests of moderate to low species richness. The distinctive flora includes Leguminosae and *Lagerstroemia* which are abundant in the canopy, and Verbenaceae including teak. The understorey comprises younger trees and saplings or thickets of bamboo. Stature varies from 5 to 40 m, depending on the underlying soil conditions, which similarly influences species richness.

Dry Dipterocarp Forest

These forests are present on acid infertile sandy and lateritic soils in continental Thailand. They are poorer in species, with a canopy dominated by a small group of fire-resistant Dipterocarpaceae, *Gluta usitata*, and in many areas by *Pinus merkusii* which is emergent and occasionally dominant. The understorey consists of saplings and tall grasses, according to frequency.

Farmland

Cultivated farmland is dominated by rice paddy in flat lowlands and level floodplains, with more diverse crops on higher ground. These crops include cassava, rubber, oil palm, coffee, and a diverse variety of fruit trees. A number of ecologically tolerant mammals can survive in the intensively cultivated farmland provided that some native trees and rough scrub or grassland remain.

Freshwater Wetlands

Not only peat swamp forest, but other freshwater wetlands are also of major importance to mammals, although their significance may have been severely reduced due to modification. The large rivers represent an important wetland habitat; but this habitat has been greatly altered through intensive human use, including both settlement and deforestation of their accessible environs. Consequently, riverine habitat and associated floodplains has rarely, if ever, been incorporated into the protected area network. The large expanses of reeds and *Saccharum* grasslands along riverine floodplains constitute other important wetland habitats, but these have been almost entirely lost, and expansive marshlands have been severely drained, canalized, and turned into paddy.

In the past 30-40 years, a number of large reservoirs have been created in steep-sided valleys for hydroelectric generation. These reservoirs support few mammals, because they are steep-sided, deep, and possess little fringing marshes.

Seagrass Beds

Seagrass beds are often found in association with mangrove, in shallow coastal waters. They form important habitat for some turtle species, which graze exclusively on the grass beds.

Zoogeography

The distribution of the world's fauna has been categorized into six zoogeographic regions. Thailand falls centrally within the Oriental Region. Within the Oriental Region, Thailand, through its geographical position, lies at the zoogeographical crossroads of Southeast Asia. Continental Thailand supports a reptilian fauna with clear affinities to Burma and India to the west, Indo-China to the east, and mixed affinities to the north. In peninsular Thailand, the fauna is predominantly Sundaic, having affinity with Malaysia and the more westerly islands of Indonesia, the Greater Sundas.

The authors have followed Deignan (1963) in dividing the country into six principal zoogeographic regions. The features of each are set out briefly below.

The North

The North is a mainly mountainous region divided by a number of north-south hilly ridges and it drains via the Rivers Ping, Wang, Yom, and Nan southward into the Chao Phraya River of Thailand's Central Plains. Small areas in the north of this region drain into the Mekong River, while parts of the west lie in the watershed of the Salween River. Virtually the entire area lies above 200 m elevation, and there are large areas of uplands above 100 m. Among the few peaks above 2000 m is Doi Inthanon, Thailand's highest mountain (2565 m). Average annual rainfall in the lowlands varies from 1000 to 1500 mm over most of the region, so that the lower elevations are dominated by deciduous forests. The climax vegetation of the higher mountains (above 800 m) is broad-leaved montane evergreen forest, sometimes referred to as hill evergreen forest. Native pines are also fairly frequent. The area of both lowland and montane forests has been much reduced and most of what is left is heavily disturbed.

The West

The West is mainly hilly, extending along the Burmese border from roughly 16°30' N to 11°40' N. Covering a great north-south span, the west is an area of great zoogeographical interest. Much of the area is hot and dry, although rainfall increases farther west and south and may be greater than 2000 mm in some areas fringing the Burmese border. The west is of great value for mammalian conservation because it supports the largest remaining expanses of forest and woody secondary growth in Thailand. The lowlands and lower hills are mainly dominated by mixed deciduous forests, including some large areas of little disturbed valley bottom along the upper reaches of the Khwae river system—the famed "River Kwai."

The Northeast

Northeast Thailand is a dry plateau at 100 to 200 m elevation; it drains into the Mekong River, which forms its northern and eastern borders. To the west, the plateau is bordered by the flat-topped mountains of the Dong Phaya Yen range and to the south, along the border with Cambodia, by the Phanom Dongrak range. The rainfall is mostly very infrequent (less than 1000 mm annually in some parts of the region's centre) and highly seasonal, so that most of the area once supported open deciduous woodlands—mainly dry dipterocarp—and savanna, with evergreen forests being largely confined to the mountain slopes. Some lowland evergreen forest formerly existed in the extreme northeast, bordering the Mekong River, but has now been almost completely cleared.

The Southeast

The Southeast is a predominantly lowland area, bordered to the south and west by the Gulf of Thailand and to the east by Cambodia. A belt of dry lowlands along the Bang Pakong River, which runs east-west, separates this region from the Northeast. The rainfall in this region is generally higher than in most other regions of continental Thailand, with some parts receiving 3000-4000 mm of rain per year; hence most of the lowlands were once covered with evergreen forest. The mountains of Khao Soi Dao, an outlier of the Cardamom mountains of Southwest Cambodia, rise to 1670 m elevation.

The Central Plains

This region comprises the alluvial basin of the Chao Phraya, which enters the Gulf of Thailand at Bangkok. Other major rivers (the Mae Klong and Tachin to the west and the Pa Sak and Bang Pakong to the east) flow through the Central Plains. Most of the area lies below 50 m elevation and some parts in the south are less than 10 m above sea level. The only significant uplands are jagged limestone outcrops and low spurs extending from hill ridges of the surrounding regions. The average annual rainfall may reach 1400 mm on the coast around Bangkok, though rainfall is substantially less (around 1100 mm) in the north of the region, toward Nakhon Sawan. The formerly extensive swamps and lowland forests of the Central Plains were mostly lost by the early 20th century as rice became a major export crop.

Peninsular Thailand

Peninsular Thailand is considered to extend southward from roughly 11°40' N latitude. The Peninsula varies in width from roughly 50 km to 220 km and a mountainous backbone runs its full length. The highest mountain is Khao Luang (1835 m) in Nakhon Si Thammarat province. The rainfall is less seasonal than in continental Thailand; annual rainfall is more than 2000 mm for most of the area and exceeds 3000 mm in some parts. Rainforest is the natural climax vegetation of the Peninsula and formerly covered almost the entire area. However, the lowlands have been all but cleared for agriculture; almost all the remaining forest lies on the steep hill slopes.

Mangrove forests are an important habitat in the Peninsula; the most extensive occur on the west coast. Sea grass beds are found in association with these forests.

The Conservation of Reptiles

This field guide describes 352 reptiles recorded within Thailand. Of these, only about one in ten species—33 species to be precise—are currently listed as threatened species in the IUCN Red List (2006), a reference database which identifies species around the world threatened with extinction.

Virtually all the turtle species of the Order Testudines have been added to the list of severely threatened animals over recent decades. Five of the six species listed as being critically endangered are turtles, namely the Leatherback *Dermochelys coriacea*, the Hawksbill Turtle *Eretmochelys imbricata*, the Narrow-headed Softshell Turtle *Chitra chitra*, the River Terrapin *Batagur baska*, and the Painted Terrapin *Batagur borneoensis*. Eight turtle species constitute all the reptile species designated as endangered: the Big-headed Turtle *Platysternon megacephalum*, the Green Turtle *Chelonia mydas*, the Olive Ridley Turtle *Lepidochelys olivacea*, the Frog-headed Giant Softshell Turtle *Pelochelys cantorii*, the Spiny Turtle *Heosemys spinosa*, the Yellow-headed Temple Turtle *Heosemys annandalii*, the Elongated Tortoise *Indotestudo elongata*, and the Asian Brown Tortoise *Manouria emys*. A further 6 turtles—the Asiatic Softshell Turtle *Amyda cartilaginea*, the Asian Box Turtle *Cuora ambionensis*, the Giant Asian Pond Turtle *Heosemys grandis*, the Malayan Snail-eating Turtle *Malayemys subtrijuga*, the Flat-shelled Turtle *Notochelys platynota*, and the Black Pond Turtle *Siebenrockiella crassicollis*, are listed as vulnerable.

The crocodilians are another Order whose members are threatened; this is another group whose members are large in size. The Siamese Crocodile *Crocodylus siamensis,* is critically endangered globally, and is represented by only a handful of individuals in Thailand. The false gharial's status is extinct in the wild(EW) for Thailand. *Crocodylus porosus* is also in high risk as critically endanger for Thailand.

The lizards and snakes number 321 species. Of these only 5 lizards are listed—all as data-deficient. These comprise the Abbott's Crested Lizard *Gonocephalus abbotti*, the Gyldenstolpe's Worm Skink *Isopachys gyldenstlopii*, the Banded Supple Skink *Lygosoma haroldyoungi,* the Khorat Supple Skink *Lygosoma koratense,* and the Scarce Ground Gecko *Dixoneus melanostictus.*

This Red List reveals two key issues: Generally speaking, the larger, more conspicuous, reptiles are severely threatened; conversely, the smaller reptiles are poorly understood in terms of identification, population sizes, and threat level, and consequently are seriously overlooked.

An unknown number of reptile species are heavily dependent upon sizable tracts of lowland forest for their survival. Unfortunately, the overwhelming majority of this vulnerable habitat was cleared for agriculture before the affected species could be protected. Consequently, it was impossible to incorporate any significantly large areas of lowland forest within the national network of protected areas.

Hunting for game meat has also had a severe impact on the remaining populations of large- and medium-sized reptiles. Selective hunting of reptilian species for body parts is of particular concern. However, the growth in international trade and global markets in recent years has placed inexorable pressure on targeted species, among which the turtles in particular seem especially vulnerable. Thailand is making great strides to curb the threat of trade and implement CITES as a full member; however it should encourage some of its neighbouring countries to make similar commitments.

Marine reptiles are in equally bad shape. All 5 sea turtle species have been designated as lower risk-conservation dependent. This highlights that the populations of marine reptiles are also highly threatened, either from disturbances on nesting beaches or from modern fishing techniques, particularly purse-seine netting.

The Protected Area System

Thailand is developing one of the best protected area systems in Southeast Asia. The kingdom has several different categories of protected areas and each serves various functions according to its enabling legislation and operating objectives. National parks—both terrestrial and marine—are lands which are gazetted to preserve and protect the natural environment, especially for the purposes of education recreation and tourism. As of December 2008, there were 73 terrestrial national parks.

In December 2008, there were also 18 marine national parks, comprising 6 archipelagos, 9 coastal parks dominated by beaches, a bay dominated by mangrove, a site dominated by diverse wetlands, and a forested site dominated by Malayan mixed dipterocarp forest. The 18 parks cover 5219 km^2.

Wildlife sanctuaries are lands preserved and protected from human activity which may disturb wildlife. Apart from one site which protects the only substantial remaining area of peat swamp forest, all the other sites encompass areas of terrestrial forest. As of December 2008, there were 55 designated wildlife sanctuaries.

While the area of natural forest and other habitats encompassed within the protected areas system is extremely large, even by international standards, the management in many of these reserves is still insufficient. To help strengthen the management of these wilderness areas, it is important that the protected area legislation is revised, and the protected area system thoroughly reviewed.

Reptilian and Conservation Bodies

The IUCN Species Survival Commission (SSC) is a science-based network of some 7000 volunteer experts from almost every country of the world, all working together toward achieving the vision of "A world that values and conserves present levels of biodiversity."

Members include researchers, government officials, wildlife veterinarians, zoo and botanical institute employees, marine biologists, protected area managers, and experts on plants, birds, mammals, fish, amphibians, reptiles, and invertebrates. Most members are deployed among the more than 100 Specialist Groups and Task Forces. Some groups address conservation issues related to particular groups of plants or animals while others focus on topical issues such as reintroduction of species into former habitats, or wildlife health.

SSC's major role is to provide information to IUCN on biodiversity conservation, the inherent value of species, their role in ecosystem health and functioning, the provision of ecosystem services, and their support of human livelihoods. SSC members also provide scientific advice to conservation organizations, government agencies, and other IUCN members, and support the implementation of multilateral environmental agreements.

Regarding reptile conservation, there are currently 4 specialist groups dedicated to the conservation of certain taxonomic groups. These are the Tortoise and Freshwater Turtle Specialist Group, the Marine Turtle Specialist Group, the Crocodile Specialist Group, and the Iguana Specialist Group.

The Tortoise and Freshwater Turtle Specialist Group (TFTSG) has initiated the Turtle Survival Alliance (TSA), an integral part of the Global Action Plan for the Conservation of Tortoises and Freshwater Turtles. The TSA is best defined as an IUCN/SSC partnership network for the sustainable captive management of freshwater turtles and tortoises. The TSA formed as a joint interdisciplinary working group of the IUCN. The TSA was a direct outgrowth of an Asian turtle workshop in Fort Worth, Texas (USA), in January 2001, titled "Developing Conservation Strategies through Captive Management."

The Marine Turtle Specialist Group (MTSG) has developed a vision: *We envision marine turtles fulfilling their ecological roles on a healthy Planet where all Peoples value and celebrate their continued survival.* The MTSG has also developed as its Mission the development and support of strategies, setting of priorities, and provision of tools that promote and guide the conservation of marine turtles and their ecological roles and habitats.

MTSG intends to accomplish this by providing advice and practical support to those who manage and conserve marine turtles; using and synthesizing the best multidisciplinary science, information, and innovative approaches; ensuring that conservation is guided by the biological constraints of marine turtles such as delayed maturity, long-distance migrations and dependence on multiple habitats; building and promoting diverse and dynamic partnerships among stakeholders, including local communities, to ensure effective networking and synergy; promoting universal appreciation of the values of marine turtles and their habitats through varied means to diverse people; building capacity to educate, inspire, and empower people to conserve marine turtles and their habitats; recognizing that marine turtles are a shared international resource; promoting conservation and population monitoring efforts that are long term in nature and anticipate environmental change; integrating local, national, and international conservation efforts; recognizing that marine turtle conservation is integrally linked to broader conservation issues.

The Crocodile Specialist Group (CSG) is a worldwide network (with 350 members) of biologists, wildlife managers, government officials, independent researchers, NGO representatives, farmers, traders, tanners, fashion leaders, and private companies actively involved in the conservation of the world's 23 living species of alligators, crocodiles, caimans, and gharials in the wild. The CSG network of experts evaluates the conservation needs of crocodilian populations, initiates research projects, conducts surveys of wild populations, estimates population numbers, provides technical information and training, and initiates conservation programs. The CSG has published Crocodiles: An Action Plan for their Conservation, and also produces the "IUCN/SSC Crocodile Specialist Group" newsletter.

In Thailand, the following organizations are involved either directly or indirectly in herpetology and the conservation of reptiles and/or their habitats

Freeland Foundation provides direct protection to wildlife in danger through wildlife law enforcement, habitat protection, education, and community outreach. The organization aims to be efficient and cost effective, implementing principles of fast intervention and empowering both government agencies and local organizations to produce results that benefit both wildlife and communities. Address: Freeland Foundation, 591 UBC II Building, 10th Floor, 1001 Sukhumvit 33, Wattana, Bangkok 10110. Tel: 02-204-2720. Fax: 02-204-2722. E-mail: info@freeland.org.

Green World Foundation promotes environmental literacy in all fields concerned with the environment. Recent projects have included publications on using biological indicators for water quality. Address: Green World Foundation, 394/46-48 Soi Penphat 4, Maharaj Road, Pranakorn, Bangkok 10200. Tel: 02-622-2250-2.

IUCN—The International Union for Conservation of Nature is assisting the Royal Thai Government with the implementation of the Convention on Biodiversity in a number of fields. This is being undertaken in collaboration with the Institute for Biodiversity Conservation. Other activities include working on water and wetlands, and working toward the undertaking of a national protected area systems plan. Address: IUCN, IUCN Asia Regional Office 63, Soi Prompong, Sukhumvit 39 Wattana, Bangkok. 10110. Thailand. Tel: 02 662 4029-33; Fax: 02 662 4387-89. E-mail: communications@iucnt.org

The National Science Museum, established in 1995 to commemorate Queen Sirikit's 60th birthday, aims to be a leader of technology and environmental issues in Asia. Address: Technopolis, Klong 5, Klong Luang, 12120. Thailand. Tel: 02 577 9999; Fax: 02 577 9991. http://www.nsm.or.th/

Naucrates, from the Greek "who dominate the sea," is an NGO based in Italy and operating in South Thailand since 1996. The organization maintains a sea turtle conservation project at Phra Thong island (Phang Nga) promoting and applying conservation strategies to avoid the extinction of the nesting population. Simultaneously, Naucrates promotes environmental education and sustainable tourim. Address: Colle Tenne–04010 Giulianello di Cori (LT) Italy; Tel: +39.3334306643; E-mail: naucrates12@tiscali.it; www.naucrates.org. Address in Thailand (from January to the end of March): P.O. Box 4, 82150 Kura Buri - Phang Nga Thailand.

The Phuket Marine Biological Centre (PMBC), contained within the Department of Marine and Coastal Resources, carries out marine science research, including fishery biology, of the coastal waters of Thailand. The knowledge will be useful for the development and conservation of living resources in these waters to the benefit of future generations. PMBC also promotes the training and education of students and scientists (both Thai and foreigners) in marine biology. Address: PMBC, Department of Marine and Coastal Resources, P.O. Box 60 Phuket, 83000. Thailand Fax: (076) 391 127. Tel: (076) 391 128. E-mail: pmbcnet@phuket.ksc.co.th

The Seub Nakhasathien Foundation promotes the conservation of the western forest complex, including the World Heritage Site of Huai Kha Khaeng and Thung Yai Wildlife Sanctuaries. Address: 693 Buillding 4, Medical Sciences Building, Yosse, Bumrung Muong Road. Pomprab, Bangkok 10100. Thailand Tel: 02 224 7838-9.

The Siam Society (Natural History Section) publishes a quality scientific journal, the *Natural History Bulletin of the Siam Society*, which appears twice each year and usually contains articles on reptiles, including behavioural studies. Address: 131 Soi 21, (Asoke) Sukhumvit Road, Bangkok 10110. Thailand Tel: 02 661 6470-75. Fax: 02 258 2491. E-mail: siams@telecom.scb.co.th.

Wetlands International is an independent international organization which aims to promote the protection and sustainable utilization of wetland resources in Asia, in conjunction with both governmental and non-governmental agencies. Address: Wetlands International—Asia Pacific, 3A39, Block A, Kelana Centre Point, SS7/19 Petaling Jaya, 47301 Selangor, Malaysia. Tel: +60-3-704 6770. Fax: +60-3-704 6772. E-mail: wiap@wiap.nasionet.net.

The Wildlife Conservation Society (WCS) was founded in 1895 as the New York Zoological Society and is based in the United States, with over 300 field projects worldwide. WCS supports comprehensive field studies to gather information on wildlife needs, train local conservation professionals, and work with in-country staff to protect and manage wildlife and wild areas for the future. Wildlife Conservation Society, P.O. Box 170, Laksi, Bangkok 10210. Tel: 02 574 0683. Fax: 02 503 4479. E-mail: wcs@wcs.org.

WWF—Thailand is an associate of the World Wide Fund for Nature (WWF) concerned with promoting the conservation of Thai wildlife through conservation education, implementing and administering projects, fundraising, and supporting the activities of the government. Address: WWF-Thailand, No. 87 Soi Paholyothin 5 (Rajakru), Paholyothin Road, Samsen nai, Phyathai, Bangkok 10400, Thailand, Tel: +662 619 8534-37 Fax: +662 619 8538-39 Email: wwfthai@wwfgreatermekong.org

Yadfon Association is a community-based conservation organization based in Trang. It focuses on mangrove and coastal resources. Yadfon Association. 16/4 Rakchan Road, Tambon Tabtieng, Amphur Muang, Trang 92000, Thailand. Tel: 075-219-737/075-214-707. Fax: 075-219-327. E-mail: yadfon@loxinfo.co.th.

Turtles
Order Testudines

This Order contains all the reptiles commonly called turtles, with their typical characteristics. It is the most primitive order of living reptiles. This group comprises shelled reptiles with solid, anapsid skulls. Worldwide this group contains 12 families, of which 6 are represented in Thailand: the big-headed turtle (Platysternidae), the leatherback turtle (Dermochelyidae), the sea turtles (Cheloniidae), the softshell turtles (Trionychidae), the freshwater pond turtles (Geoemydidae), and the tortoises (Testudinidae).

KEY TO THE THAI TESTUDINES

1 a	Carapace and plastron covered with clearly demarcated horny shields	**2**
b	Carapace and plastron covered with continuous layer of undivided skin	**5**
2 a	Front limbs in form of elongate flippers, without separated digits; each forelimb with one or two claws; intergular present; marine	*Cheloniidae*
b	Front limbs club-shaped or with distinct digits, each limb bearing four or five claws; intergular present or absent	**3**
3 a	Inframarginal scute(s) present, separating pectoral and abdominal scutes from marginal scutes; tail as long as shell	*Platysternidae*
b	Inframarginal scute absent; pectoral and abdominal scutes in contact with marginal scutes	**4**
4 a	Hind feet elephantine, front limbs usually with osteoderms; terrestrial	*Testudinidae*
b	Hind feet not elephantine, but usually webbed, no osteoderms on forelimbs; typically aquatic	*Geoemydidae*
5 a	Carapace raised into five or seven prominent longitudinal ridges, upper jaw strongly cusped; limbs clawless	*Dermochelyidae*
b	Carapace never ridged; upper jaw not cusped, claw present, neck long, snout elongate	*Trionychidae*

Key modified from Dollinger (1985).

Turtles Family Platysternidae

Member of this family is aquatic. The limbs are not paddle-shaped. The digits are distinct with four or five claws. The shell is covered with horny shields. The plastral shields are separated from the marginals by inframarginals. The tail is as long as the body. One species occurs in Thailand.

1 Big-headed Turtle *Platysternon megacephalum*

Carapace Length: Up to 18.4 cm.

This species has a diagnostic huge triangular head, so large than it cannot be withdrawn into the carapace. The head, which is covered with horny plates on the top and sides, ranges from yellow-brown to olive, and may be marked with dark brown, orange, or red longitudinal striations. The well-developed jaws are covered with horny plates. The comparatively small carapace is barely 20 cm wide, being yellow brown to olive and may contain some orange or red blotches. The plastron and bridge are yellow. The tail is as long as the body, or longer. **Behaviour:** Mainly nocturnal, spending the daytime burrowed into gravel, in piles of submerged leaves, or under rocks. At night or during heavy rain, it wanders out of the water searching for food along the streams and in streamside vegetation. Probably carnivorous, feeding on small fish and invertebrates, especially freshwater crabs, on stream bottoms. It climbs occasionally, both up inclined tree trunks and over stones and boulders. Females lay 1-2 elliptical eggs per clutch on an exposed sandbar. **Habitat:** Cool, mountain streams in lower and upper montane forest. Occasionally individuals are found at lower elevations, if they have been swept downstream by strong currents. **Range:** Northern, northeastern, and western Thailand. **Regional Range:** Myanmar, China, Vietnam, Laos, Cambodia, and Thailand. **International Conservation Status:** IUCN: Endangered; CITES: Appendix II.

1

Leatherback Turtle Family Dermochelyidae

This family is represented by a single species, *Dermochelys coriacea*, which is an extremely large pelagic species.

2 Leatherback Turtle *Dermochelys coriacea*

Carapace Length: Up to 244 cm.

The largest and heaviest of all living turtles and tortoises. The elongated, lyre-shaped carapace is smooth-skinned and leathery, almost rubber-like. It ranges in colour from brown to black. There are no epidermal shields but there are seven prominent longitudinal ridges. There is no sharp angle between the carapace and plastron. The head and neck are black or brown, with white or yellow blotches. The snout is blunt and non-projecting. The limbs are paddle-shaped and clawless, while the feet lack scales. Tails of males are longer than their hind limbs; female tails are much shorter. **Behaviour:** Pelagic, though occasionally enters bays or estuaries. Omnivorous diet, preferring coelenterates and cephalopods; also feeds on sea urchins, squid, molluscs, crustaceans, and fish. It also eats floating seaweed. Nests on traditional beaches at night, usually on the mainland, 3-7 times per year. Known nesting sites are on the deep shores of Phang-nga and Phuket provinces. Clutch size is 50–170 eggs. **Habitat:** Marine. **Range:** Found in the Andaman Sea and the Gulf of Thailand. **Regional Range:** This giant ranges throughout the warm and temperate waters of the Atlantic, Indian, and Pacific oceans. **International Conservation Status:** IUCN: Critically endangered; CITES: Appendix I.

2

Sea Turtles Family Cheloniidae

Members of this family are marine. The limbs are paddle-shaped to facilitate strong swimming in the seas and oceans. They have one or two claws.

The shell is covered with horny shields. Worldwide: 4 living genera and 6 species. Thailand: 4 genera and 4 species.

KEY TO THE THAI CHELONIIDAE

1 a Carapace with four lateral scutes on each side, the first pair not in contact with the precentral scute **2**

b Carapace with five lateral scutes on each side, the first pair in contact with the precentral scute **3**

2 a Carapace elliptical, covered by imbricate scutes in very old individuals; head narrow, with two pairs of prefrontal scales; jaw hawk-like, not serrated; flippers usually with two evident claws *Eretmochelys imbricata*

b Carapace nearly oval, with no imbricate scutes; head blunt (short snout), the preorbital distance clearly smaller than orbital length; a single pair of prefrontal scales, usually four postorbital scales; lower jaw serrated; flippers usually with only one distinct claw *Chelonia mydas*

3 a Carapace cordiform, its length always greater than the width; plastron usually with three pairs of inframarginal scutes; generally without pores; carapace scutes thick and rough to touch; head comparatively large, with a heavy and strong lower jaw lacking an internal alveolar rim; body colour usually reddish-brown or yellowish-brown *Caretta caretta*

b Carapace nearly round, its length similar to the width; plastron usually with four pairs of pored inframarginal scutes; lateral scutes are often more than five pairs; carapace scutes smooth to touch; head moderately small, with a cutting lower jaw provided with an internal alveolar rim; fore flippers with one or two visible claws on anterior border, sometimes another claw on distal part; rear flippers with two claws; body colour grey-olive or olive-yellowish *Lepidochelys olivacea*

Key modified from Marquez (1990).

Green Turtles Genus *Chelonia*

This genus is found in the Atlantic, Indian, and Pacific oceans, mainly in the tropics. The lower jaw is strongly serrated. It has four pairs of pleurals. Worldwide: 2 species. Thailand: 1 species.

3 **Green Turtle** *Chelonia mydas*

Carapace Length: Up to 153 cm.

A medium- to large-sized sea turtle. The carapace is oval to heart-shaped, and is serrated posteriorly. All skin is brown or occasionally black to grey, and many head scales may have yellow margins. It has four pairs of pleurals (like the Hawksbill Turtle); by comparison the Loggerhead Turtle and the

Olive Ridley Turtle have five pairs. The lower jaw is strongly serrated; in the Hawksbill Turtle the jaw is only weakly serrated. **Behaviour:** Omnivorous, feeding on crustaceans, small molluscs, jellyfish, and sponges as well as algae, mangrove roots and leaves, and seagrass. Nesting is nocturnal. The female nests in 2- or 3- year cycles, nesting several times each season, 10-15 days apart. **Habitat:** Marine. **Range:** Found in the Andaman Sea and the Gulf of Thailand, including Khram Island and Kra Island. **Regional Range:** Occurs throughout the warm waters of the Atlantic, Indian, and Pacific oceans. **International Conservation Status:** IUCN: Endangered; CITES: Appendix I.

3

Ridley Turtles Genus *Lepidochelys*

Members of this genus have heart-shaped carapaces with serrated posterior marginals. They have 5-8 pleurals on each side. Some individuals have more than five vertebral scutes. Worldwide: 2 species. Thailand: 1 species.

4 **Olive Ridley Turtle** *Lepidochelys olivacea*

Carapace Length: Up to 71.0 cm.

Smaller than most other sea turtles with 6-8 pleurals on each side. The heart-shaped olive carapace is flattened dorsally and serrated posteriorly. The plastron and bridge are greenish-white or greenish-yellow. The skin is olive above and lighter below. **Behaviour:** Highly carnivorous, feeding on fish, crabs, snails, urchins, and jellyfish. Seaweed is occasionally consumed. Mating probably occurs offshore of the nesting beaches during the nesting period. Females usually nest twice in a season, rarely three times, often during periods of high winds. **Habitat:** Usually occurs in shallow seas, large bays, lagoons, and in the lea of reefs, though occasionally strays into deep marine waters. **Range:** Found in the Andaman Sea and the Gulf of Thailand. **Regional Range:** Occurs throughout the Indian and Pacific oceans. **International Conservation Status:** IUCN: Endangered; CITES: Appendix I.

4

Loggerhead Turtle Genus *Caretta*

This genus is found in the Atlantic, Indian, and Pacific oceans. It wanders widely. Worldwide: 1 species. Thailand: 1 species.

5 **Loggerhead Sea Turtle** *Caretta caretta*

Carapace Length: Up to 213 cm.

One of the larger sea turtles, with five or more pleurals. The elongated carapace has a medial vertebral keel that becomes progressively smoother with age and is serrated posteriorly. The carapace is reddish brown but may be tinged olive. The plastron and bridge are yellow or cream-coloured. The head ranges from reddish- or chestnut-yellow to olive-brown, often with yellow-bordered scales. Males have narrow shells and long thick tails. **Behaviour:** Omnivorous, feeding on sponges, sea urchins, squid, molluscs, oysters, barnacles, and crustaceans and fish. It also eats seaweed and sea grass. Nesting does not occurs in Thailand. Both 2- and 3-year cycles occur. **Habitat:** Wanders widely through marine waters; it enters bays and estuaries. **Range:** Vagrant to the Andaman Sea and the Gulf of Thailand. Recorded in Rayong, Ranong, Phuket, and Songkhla provinces. **Regional Range:** Occurs widely throughout the Atlantic, Indian, and Pacific oceans. **International Conservation Status:** IUCN: Insufficiently Known; CITES: Appendix II.

5

Hawksbill Turtle Genus *Eretmochelys*

This genus is tropical and aquatic. All eight pairs of costals are prevented from touching medially. Worldwide: 1 species. Thailand: 1 species.

6 **Hawksbill Turtle** *Eretmochelys imbricate*

Carapace Length: Up to 91.0 cm.

A small- to medium-sized turtle with four pairs of pleurals. The carapace is shield-shaped and dark greenish brown. The plastron is hingeless and yellow. The head scales are black or chestnut brown at the centre and lighter at the margins. The jaws are yellow with some brown streaks. The chin and throat are yellow. Males have slightly concave plastrons, long tails, and long heavy claws. **Behaviour:** Omnivorous, preferring marine invertebrates such as sponges, coelenterates, sea urchins, molluscs, barnacles, and crustaceans and fish. It also eats fruit, leaves, bark, and wood of mangrove species. Mating occurs in shallow waters off the nesting beaches. Nesting occurs at night in three-year cycles. Some 53-206 eggs are laid in each nest. **Habitat:** Characteristically found in rocky places and reefs. Also frequents shallow coastal waters, mangrove-fringed bays, and estuaries. **Range:** Found in the Andaman Sea and the Gulf of Thailand, including the islands of Koh Khram and Koh Kra. **Regional Range:** Occurs throughout the Atlantic, Indian, and Pacific oceans. **International Conservation Status:** IUCN: Critically Endangered; CITES: Appendix I.

6

Softshell Turtles Family Trionychidae

Members of this family are aquatic, freshwater, with smooth skin completely covering the shell. The digits are distinct and they have three claws. Five species are known in four genera from Thailand.

KEY TO THE THAI TRIONYCHIDAE

1	**a**	Snout (proboscis) as long as greatest diameter of orbit	**2**
	b	Snout shorter than greatest diameter of orbit	*Pelochelys*
2	**a**	Postorbital arch much broader than greatest diameter of orbit; triturating surfaces narrow and sharp	*Chitra*
	b	Postorbital arch narrower than greatest diameter of orbit; triturating surfaces broad and flat	**3**
3	**a**	All eight pairs of costals prevented from touching medially by neural bones; juvenile with red spot behind the eye	*Dogania*
	b	7th or 8th pair of costals touch medially, not separated by neural; juvenile without red spot behind the eye, head olive with yellow spots; five plastral callosities	*Amyda*

Key modified from Ernst and Barbour (1989).

Common Softshell Turtle Genus *Amyda*

This genus is tropical and aquatic. It has five plastral callosities. They have long noses. The nuchal is at least three times as broad as it is long. Worldwide: 1 species. Thailand: 1 species.

7 Asiatic Softshell Turtle *Amyda cartilaginea*

Carapace Length: Up to 70.0 cm.

Distinguished from other softshell turtles by having numerous small yellow spots on the olive-coloured head, neck, and limbs. Usually there are large orange or pink markings on the sides of the head. The oval carapace is olive-grey to greenish-brown with numerous yellow-bordered black spots. The male plastron is white while the female plastron is greyish. **Behaviour:** Almost entirely aquatic. Mainly nocturnal. It has a carnivorous diet, feeding on fish, amphibians, crustaceans, aquatic insects, and other aquatic invertebrates. Annually lays 3 to 4 clutches of 3-6 eggs. **Habitat:** Occurs in slow-flowing rivers and streams in low-lying areas as well as upland waterways. Also found in marshes and ponds. **Range:** Found throughout Thailand. Recorded in Kamphaeng Phet, Uthai Thani, Sa Kaeo, Chanthaburi, Chonburi, Pathum Thani, Surat Thani, and Satun provinces. **Regional Range:** Myanmar, Vietnam, Laos, Cambodia, Thailand, Malaysia, and Indonesia. **International Conservation Status:** IUCN: Vulnerable; CITES: Appendix II.

7

Narrow-Headed Softshell Turtles Genus *Chitra*

This group is tropical and aquatic. They have narrow heads—the triturating surface is narrow and sharp. They have short noses. Worldwide: 3 species. Thailand: 2 species.

KEY TO THE THAI *CHITRA*

1 a Overall coloration greenish-yellow to olive-green; carapacial pattern simple; paramedian neck stripes form a bell-like pattern on anterior carapace and neck stripes that form a continuous rim around carapace; lacks a distinct third pair of neck stripes *C.chitra*

b Overall coloration chocolate brown; carapacial pattern moderately complex; paramedian neck stripes absent on anterior carapace and neck stripes do not form a continuous rim around carapace; distinct third pair of neck stripes *C.vandijki*

Key modified from McCord and Pritchard (2002).

8 **Narrow-headed Softshell Turtle** *Chitra chitra*

Carapace Length: Up to 110.0 cm.

A large softshell with a flattened rounded carapace which is pale-brown to olive with irregular patterning of darker-edged, brown blotches; by comparison, the Myanmar Narrow-headed Softshell Turtle has a chocolate-brown body coloration. The carapacial pattern is simple with a carapacial midline usually present. The neck stripes form a bell-like pattern on the anterior carapace; the neck stripes also form a continuous light rim around the carapace; no such rim is present in Myanmar Narrow-headed Softshell Turtle. The small blunt head and a telescopic proboscis-like snout are olivaceous with characteristic dark-bordered yellow streaks. **Behaviour:** Almost entirely aquatic, only coming onto river beach to lay eggs. Mainly nocturnal. It has a mainly carnivorous diet, feeding on fish, molluscs, crustaceans, and other aquatic invertebrates; on occasion it eats aquatic plants. In the daytime it buries itself into the sandy substrate. Female produces 60-110 eggs per clutch. **Habitat:** Occurs in large sandy sections of slow-flowing rivers. **Range:** Occurs in the MeKlong drainage system in Ratchaburi, Kanchanaburi, and

Tak provinces, in western Thailand. **Regional Range:** Thailand, Peninsular Malaysia and Indonesia (Java and Sumatra). **International Conservation Status:** IUCN: Critically Endangered; CITES: Appendix II. It is one of the most endangered turtle species in the world.

8

9 Myanmar Narrow-headed Softshell Turtle *Chitra vandijki*

Carapace Length: Up to 38.5 cm.

A large softshell turtle with an overall brown coloration. It has a characteristic "V" neck marking. The neck also has three pairs of black-bordered neck stripes running parallel to the median neck stripe, giving a total of seven neck stripes. By comparison the Striped Giant Softshell Turtle has a total of five neck stripes. It lacks both a midline carapacial stripe and a continuous light rim around the carapace. The plastron is pinkish or white. **Habitat:** Occurs in slow-flowing rivers. **Range:** Recorded from the Salween River in Tak Province, western Thailand. **Regional Range**: Myanmar and Thailand.

9

Hillstream Softshell Turtle Genus *Dogania*

This genus is tropical and aquatic. All eight pairs of costals are prevented from touching medially. Worldwide: 1 species. Thailand: 1 species.

10 **Hillstream Softshell Turtle** *Dogania subplana*

Carapace Length: Up to 19.4 cm.

A medium-sized softshell turtle with a very flat, oval carapace. The carapace ranges from olive to black or dark brown; it has two or three pairs of black-centered ocelli with yellow outer rings. The plastron is whitish to cream or grey, and bears four weakly developed callosities. The head is coloured olive to brown or black with some yellow or black spots. A black medial stripe may pass along the top of the snout. The male has a longer thicker tail than the female. **Behaviour:** Aquatic. Adults keep to deeper, more turbid water, burying themselves in mud or under piles of submerged vegetation. Young animals often hide beneath rocks and large stones. **Habitat:** Inhabits clean hill streams in dry evergreen forest, mixed deciduous forest, peninsular monsoonal evergreen forest, and Malayan mixed dipterocarp forest. **Range:** Western and peninsular Thailand. **Regional Range:** Myanmar, Thailand, Malaysia, Indonesia, and the Philippines. **International Conservation Status:** IUCN: Insufficiently Known; CITES: Appendix II.

10

Giant Softshell Turtles Genus *Pelochelys*

This group is tropical and aquatic. The snout is shorter than the greatest diameter of the orbit. Worldwide: 2 species. Thailand: 1 species.

11 **Frog-headed Giant Softshell Turtle** *Pelochelys cantori*

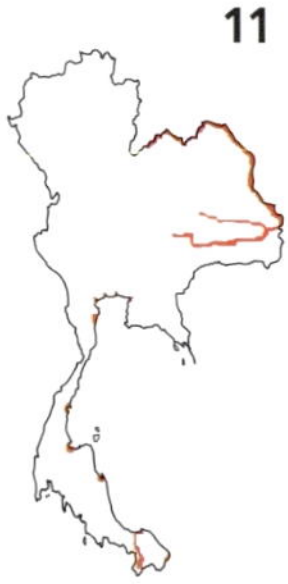

Carapace Length: Up to 129.0 cm.

The largest living softshell turtle, with eyes located very close to the tip of the short snout. The head is olive with small dark spots; the neck and limbs are olive dorsally, cream below. The flat, rounded carapace is olivaceous to brown with minute black spots, or lines extending outward from the vertebral area. The plastron is cream-coloured. The smaller males have shorter, thicker tails than females. **Behaviour:** Mainly aquatic, spending long periods underwater. Hides under overhanging rocks or by burying itself in the sandy, mud substrate. It has an omnivorous diet, consuming fish, shrimps, crabs, molluscs and aquatic plants. **Habitat:** Found in deep, slow-moving rivers and streams, often far inland. Also found in brackish estuaries. **Range:** Northeastern, central, and peninsular Thailand. Recorded in the Mekong Basin, Bangkok, Petchaburi and Yala provinces. **Regional Range:** India, Bangladesh, Myanmar, China, Vietnam, Laos, Cambodia, Thailand, Malaysia, Indonesia, the Philippines, and Papua New Guinea. **International Conservation Status:** IUCN: Endangered; CITES: Appendix II.

11

Freshwater Pond Turtles Family Geoemydidae

This family comprises hard-shelled, amphibious, herbivorous, or carnivorous and omnivorous chelonians. The majority are relatively small-sized, inhabiting the various freshwater bodies. They are more adapted to an aquatic freshwater life than to a terrestrial one; the limbs are more or less flattened, the digits are webbed and the top of the head is either covered with smooth skin or with the posterior part of it divided into shields. They lack mesoplastron infralabials and gular shields. They are closely related to the land tortoises which have elephantine hind feet. Fifteen species are known from Thailand.

KEY TO THE THAI GEOEMYDIDAE

1 a	Plastron with freely movable hinge between pectoral and abdominal scutes	**2**
b	Plastron rigid, lacking a hinge, or with only a partially movable hinge in adult females	**5**
2 a	Posterior margin of carapace unserrated; plastal lobes wide, capable of closing shell completely	*Cuora*
b	Posterior margin of carapace serrated; plastral lobes narrow, not capable of closing shell completely	**3**
3 a	Five vertebral scutes	*Cyclemys*
b	Six of seven vertebral scutes	*Notochelys*
4 a	Top of head lacks orange markings or spots	**5**
b	Top of head has orange markings or spots	*Heosemys-part*
5 a	Triturating surface of maxilla moderate to broad posteriorly, but medially narrowed	*Siebenrockiella*
b	Triturating surface of maxilla broad throughout its length	**6**
6 a	Snout projecting and somewhat turned up at the end; carapace of adults smooth	*Batagur*
b	Snout not turned up at the end; carapace of adults with 1-3 longitudinal keels	**7**
7 a	Plastron with brown radial pattern; carapace with only a single longitudinal keel	*Heosemys-part*
b	Plastron lacks radial pattern; carapace with 3 longitudinal keels	*Malayemys*

Key modified from Ernst and Barbour (1989).

Mangrove Terrapins Genus *Batagur*

This group is represented by two river-inhabiting species, which are among the largest geoemydid turtles. It has four or five claws on the forefoot. The triturating surface of the maxilla has one or two distinct median ridges.

12 **Mangrove Terrapins** *Batagur baska*

Carapace Length: Up to 61.0 cm.

One of the largest aquatic terrapins, reaching over 60 cm in length. Characterized by having broad, webbed, front feet bearing four claws; in other species five claws are present. The somewhat domed shell is uniformly brownish-olive or greenish-grey. The plastron and broad bridge are uniformly cream or yellow. The head is comparatively small, with an upturned, pointed, and strongly projecting snout. The head coloration shows variation depending upon the breeding condition, with both sexes usually being olive-grey above, and lighter grey laterally and ventrally. In the breeding season the head, neck, and legs of both sexes turn black; and the iris of the male changes from creamy yellow to white. The base of the neck is rich crimson. The forelimbs are paddle-shaped. Males have thicker, longer tails than females. **Behaviour:** Has a shy disposition. Omnivorous, although prefers a herbivorous diet of stems, leaves, and mangrove fruits of river-edge plants. May also eats molluscs, crustaceans and fish. Nests in colonies on sandbars. Lays two or three clutches of about 20 eggs each over a 6-week period. **Habitat:** Favoured habitat is shallow, muddy tidal regions at wide river mouths, lined with mangrove and other vegetation. During the breeding season the females migrate far upstream to nest. **Range:** Restricted to lower peninsular Thailand. Recorded from Ranong, Pattani, and Satun provinces. **Regional Range:** Bangladesh, Myanmar, Vietnam, Cambodia, Thailand, Malaysia and Indonesia (Sumatra). **International Conservation Status:** IUCN: Critically Endangered; CITES: Appendix I.

12

13 Painted Terrapins *Batagur borneoensis*

Carapace Length: Up to 76.0 cm.

Probably the largest geoemydid turtle, very occasionally reaching more than 70 cm in length, though most individuals currently are under 60 cm. Very similar to the Mangrove Terrapin but has five claws on the webbed front feet; the Mangrove Terrapin has four. The oval carapace is light-brown to olive in the adults, with three broad black longitudinal stripes. The plastron is uniformly cream or yellow. The head of the non-breeding male is grey, while the female's head is olive-green. During the breeding season the male develops a brilliant white head, and a red forehead edged with black. Females grow much larger than males. **Behaviour:** Feeds on fruits, leaves, and clams. Females migrate up river to lay one to three clutches of 15-25 eggs on sandy spits. **Habitat:** Found in mangroves and estuaries. **Range:** Peninsular Thailand, as far north as Bangkok. Recorded in Bangkok and Satun Province. **Regional Range:** Thailand, Malaysia, and Indonesia (Sumatra and Borneo). **International Conservation Status:** IUCN: Critically Endangered; CITES: Appendix II.

13

Southeast Asian Box Turtles Genus *Cuora*

This group is tropical and generally semi-aquatic. The plastral lobes are wide which are capable of closing the shell completely. Worldwide: 9 species. Thailand: 1 species.

14 **Asian Box Turtle** *Cuora amboinensis*

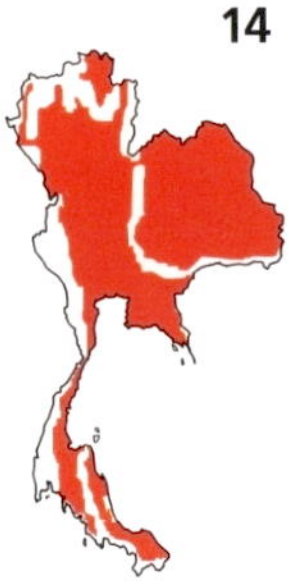

Carapace Length: Up to 21.0 cm.

A small-sized turtle. The head shows characteristic yellow striping; one stripe runs posteriorly from the tip of the protruding snout through the orbit to the neck, while two other yellow stripes are also present on the blackish sides of the head. Dorsally the head is olive or dark-brown, and yellow below. The tympanum is olive or brown. The strongly domed carapace is uniformly dark-olive, brown or black. The plastron, which has a complete hinge throughout life, is yellow to light brown, with the outside of each scute having a large dark brown or black spot. Males have slightly concave plastrons. Limbs are olivaceous or yellowish. **Behaviour:** Largely aquatic, although occasionally adults may be found far from water. Has quite a retiring nature. Largely herbivorous diet, feeding on fruits and other soft vegetation, as well as earthworms and other small prey. Lays clutches of 2-4 elongate eggs. **Habitat:** Favours lowland wetlands, including marshes, ponds, pools in streams, and flooded rice paddies. **Range:** Found in lowlands throughout Thailand. Recorded in Chiang Mai, Bangkok, Surat Thani, and Phang-nga provinces. **Regional Range:** India, Bangladesh, Myanmar, Vietnam, Laos, Cambodia, Thailand, Malaysia, Indonesia (Sulawesi), and the Philippines. **International Conservation Status:** IUCN: Vulnerable; CITES: Appendix II.

14

Asian Leaf Turtles Genus *Cyclemys*

This group is tropical and generally semi-aquatic. The carapace is only slightly flattened across the vertebrals, and is serrated posteriorly. It has five vertebral scutes. Worldwide: 7 species. Thailand: 4 species.

KEY TO THE THAI *CYCLEMYS*

1 a	Plastron in adult is dark-brown to black	**2**
b	Plastron in adult mainly yellow	**3**
2 a	Crown of the head speckled; all juveniles and adults from the eastern part of the range with conspicuous head and neck stripes, adults from the western part of the range unstriped	*C.oldhamii*
b	Crown of the head uniformly copper or light brown, head and neck unstriped	*C.enigmatica*
3 a	Femoral mid-seam shorter than anal mid-seam; anal notch small and acute-angled; throat striped or with light and dark vermiculations, light head and neck stripes narrow (in live specimens reddish); in aged individuals bridge never covered by massive black bar	*C. dentata*
b	Femoral mid-seam shorter than or equal to anal mid-seam; anal notch small to wide and acute-angled to obtuse-angled; throat uniformly light coloured, light head and neck stripes wide (in live specimens yellow or salmon); in aged individuals bridge covered by massive black bar	*C. atripons*

Key modified from Iskandar (2000), Iverson & McCord (1997) and (Fritz et al 2008).

15 Asian Leaf Turtle *Cyclemys dentata*

Carapace Length: Up to 24.0 cm.

A smaller-sized leaf turtle. Distinguished from other leaf turtles by having a uniformly or mostly yellow plastron. The moderately large head and neck is striped, sometimes speckled with fine black spots on top. The carapace is fairly depressed, being dark-brown or olive brown or black. In juveniles the shell is tinged with chestnut and may have dark rays present. The posterior margin of the carapace is strongly serrated in the juvenile, but less so in the adult; this feature is common to all members of the genus. The snout is short, slightly projecting beyond lower jaw. The limbs are dark brown. **Behaviour:** Omnivorous, feeding on fruits, leafy vegetation, invertebrates and dead animals. Lays 1-3 hard-shelled eggs per clutch. **Habitat:** One of the most common turtles of forested streams and environs. Juveniles are found in hill streams, while adults are more terrestrial. Found in a range of forest habitats, at altitudes between 300 and 1,500 metres. **Range:** Found in peninsular Thailand, south of the Isthmus of Kra. **Regional Range:** Thailand, Malaysia, Indonesia and the Philippines. **International Conservation Status:** IUCN: Lower Risk; CITES: Appendix I.

Plastron

15

16 Black-striped Leaf Turtle *Cyclemys oldhamii*

Carapace Length: Up to 25.0 cm.

A relatively large *Cyclemys* species, resembling other black-bellied leaf turtles. The carapace, which has a single keel, and the soft parts of the body are dark brown. The head and neck are dark and faintly striped, especially in juveniles and turtles from the eastern part of the range (formerly known as *C. tcheponensis*). The plastron is dark with boldly radiating dark lines. Juveniles are uniformly dark-coloured or with small dark dots, with the bold lines soon appearing with growth. Forelimbs have five claws, the hind limbs have four. **Behaviour:** Omnivorous, feeding on fruits, leafy vegetation, invertebrates and dead animals. Lays 1-3 hard-shelled eggs per clutch. **Habitat:** Found in shallow stretches of rivers in plains and lower hills.

Usually hides in vegetation during the daytime. **Range:** Found in forested areas throught the country. **Regional Range:** India, Bangladesh, Myanmar, Vietnam, Laos, Cambodia and Thailand. **International Conservation Status:** Presently unlisted.

Plastron

16

17 Dark Leaf Turtle *Cyclemys enigmatica*

Carapace Length: Up to 22.0 cm.

Medium-sized leaf turtle, similar to Black-striped Leaf Turtle, but the head and neck are unstriped and the crown of the head is light uniform copper. The plastron is dark with boldly radiating dark lines. Juveniles are uniformly dark-coloured or with small dark dots, with the bold lines soon appearing with growth. The forelimbs have five claws, the hind limbs have four. **Behaviour:** Semi-aquatic. Omnivorous. Agile and lively. **Habitat:** Found in Malayan mixed dipterocarp forest. Lives in streams and streamlets but also spends considerable time on the forest floor. **Range:** Found in Hala Bala, Yala Province in the extreme south in peninsular Thailand. **Regional Range:** Thailand, Malaysia, and Indonesia.

17

Plastron

18 Black-bridged Leaf Turtle *Cyclemys atripons*

Carapace Length: Up to 22.4 cm.

A medium-sized leaf turtle. The head is mottled with black dorsally and has a cream-to-brown temporal and post-orbital stripe on each side of the head. The chin is nearly immaculate cream or yellow. The carapace pattern is light olive to brown to nearly black, with coarse rays radiating from the areolae. The plastron is uniformly yellow though may have a few coarse black radiating lines. The bridge is strongly streaked with black or a solid black bar. **Behaviour:** Semi-aquatic. Omnivorous. **Habitat:** Confined to mountainous areas in southeastern Thailand. Found in southeastern monsoonal evergreen forest. **Range:** Found in southeastern Thailand, including Koh Chang and Koh Kut. Recorded in Khao Ang Ru Nai, Chachaengsao Province; Khao Khieo-Khao Chompu, Chonburi Province; and Khao Kuap, Trat Province. **Regional Range:** Cambodia and Thailand.

Plastron

18

Asian Pond Turtles Genus *Heosemys*

This Asian tropical group is semi-terrestrial or semi-aquatic. The carapace is usually flattened across the vertebrals Worldwide: 6 species. Thailand: 3 species.

KEY TO THE THAI *HEOSEMYS*

1	**a**	Anterior margins of carapace not strongly serrated	**2**
	b	Anterior margins of carapace strongly serrated	*H. spinosa*
2	**a**	Orange spots on top of head	*H. grandis*
	b	Yellow spots on top of head	*H. annandalei*

Key modified from Taylor (1970).

19 Giant Asian Pond Turtle *Heosemys grandis*

Carapace Length: Up to 43.5 cm.

One of the largest hard-shelled Asian turtles. Moderately sized head is greyish green with yellow or orange speckling which disappears with age. The broad, oval carapace which is strongly domed and only serrated along posterior margin, is brown to greyish-brown or black. A pale streak may occur along the keel. Light seams may appear between the scutes. The large plastron, bridge, and undersides of the marginals are pale yellow; in the juvenile these features are brighter yellow with striking black radiating lines. In the longer, thicker-tailed males the plastron is concave; in females it is flat. **Behaviour:** Semi-aquatic, spending much time out of water in concealment. **Habitat:** Inhabits a wide range of wetlands, including rivers, streams, swamps, lakes, and marshes over a range of altitudes. **Range:** Found in western and peninsular Thailand. Recorded from Uthai Thani, Bangkok, Chumphon, Narathiwat, and Yala provinces. **Regional Range:** Myanmar, Vietnam, Laos, Cambodia, Thailand, and West Malaysia. **International Conservation Status:** IUCN: Vulnerable; CITES: Appendix II.

19

20 Spiny Turtle *Heosemys spinosa*

Carapace Length: Up to 22.0 cm.

The carapace is much depressed, and brown with a light streak along the median keel. The juvenile has a characteristic strongly serrated carapace, with each marginal having a pointed outer edge; small spines are also found at the posterior margin of each plural scute. With age these spines get worn down such that mature individuals may have a markedly smoother carapace. The rather small head is brown or greyish with a yellow spot near the tympanum. The neck is reddish, sometimes with longitudinal yellow stripes. The large plastron and underside of the marginals are yellow with characteristic radiating dark lines on each scute. The limbs are grey with some yellow spotting. The toes are partially webbed. Like many species, the males have slightly longer, thicker tails than females. **Behaviour:** Semi-aquatic, in cool, shady areas near streams. Often concealed in leaf litter, under tree roots, or in tree holes not far from streams. During nesting, 1-2 very large eggs are laid several times each year. **Habitat:** Shallow, clear mountainous streams in peninsular monsoonal evergreen forest and Malayan mixed dipterocarp forest. **Range:** Found in peninsular Thailand. **Regional Range:** Myanmar, Thailand, Malaysia, Singapore, and Indonesia. **International Conservation Status:** IUCN: Endangered; CITES: Appendix II.

20

21 Yellow-headed Temple Turtle *Heosemys annandalii*

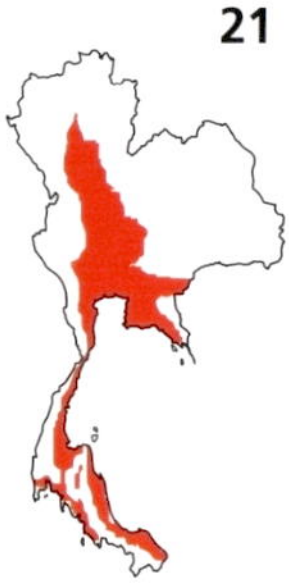

Carapace Length: Up to 60.0 cm.

A large-sized turtle with an elongated carapace. The carapace is serrated posteriorly, being uniformly dark brown or black. The moderate-sized head is blackish to olive with dense yellow speckling. The plastron and bridge are yellow with large black blotches on each scute. The juvenile is very different, having a broad rounded carapace with a sharp vertebral keel. It also has a black head and neck with distinct yellow stripes. The limbs and short tail are grey or greenish. **Behaviour:** Mainly aquatic; adults are always seen in the water or floating among vegetation, while the young are partially terrestrial. Herbivorous, feeding on aquatic plants, fruits. Lays four elongated eggs. **Habitat:** Occurs in slow-moving rivers, swamps, and flooded rice paddies. Often found in temple ponds. **Range:** Uncommon. Found in the Central Plains and peninsular Thailand. Recorded in Bangkok as well as Pathum Thani, Samut Prakan, and Pattani provinces. **Regional Range:** Vietnam, Laos, Cambodia, Thailand, and West Malaysia. **International Conservation Status:** IUCN: Endangered; CITES: Appendix II.

21

Snail-Eating Turtles Genus *Malayemys*

This genus is aquatic. The carapace is unserrated posteriorly. It has three longitudinal keels. The ridge on the triturating surface of the maxilla is reduced and indistinct. Worldwide: 2 species. Thailand: 2 species.

KEY TO THE THAI *MALAYEMYS*

1 a	Front of upper jaw with 1-2 pairs of vertical white lines	*M. macrocephala*
b	Front of upper jaw with three vertical lines	*M. subtrijuga*

Key prepared by Chan-ard, Parr and Nabithabhata.

22 Malayan Snail-eating Turtle *Malayemys macrocephala*

Carapace Length: Up to 21.0 cm.

Small to medium-sized aquatic turtle with a relatively large, black-coloured head with several cream, white, or yellow stripes. The head pattern comprises a horizontal ivory stripe passing from the tip of the snout, above the eye and ear to the back of the mouth, and runs back farther discontinuously; in the Indo-chinese Snail-eating Turtle this stripe is broken behind the eye. A second ivory stripe runs from the side of the snout below the eye to the back of the head. This second stripe does not meet with the first stripe; in the Indo-chinese Snail-eating Turtle this second stripe has a connecting line to the first stripe. The tip of the snout is ivory white, with only two vertical stripes. The pupil lacks ivory rings. The moderately arched, oval carapace ranges from light to dark brown with a characteristic cream or yellow border, and lacks posterior serrations. The yellow or cream plastron has large dark brown or black blotches on each scute, lacking any scute rings. Two dark marks are also found on the bridge. The limbs are grey to black with a narrow yellow outer border. The tail is very

22

short. **Behaviour:** Nocturnal. Mainly carnivorous, having a specialized diet of snails, although also eat worms, aquatic insects, crustaceans, and small fish. **Habitat:** Slow-moving waterbodies, including small streams, canals, ponds, marshes, and rice paddies. **Range:** Fairly abundant, being found in lowland areas in all regions of the country apart from the northeast. **Regional Range:** Thailand and Malaysia. **International Conservation Status:** IUCN: Vulnerable; CITES: Appendix II.

23 Indo-chinese Snail-eating Turtle *Malayemys subtrijuga*

Carapace Length: Up to 21.0 cm.

Small to medium-sized aquatic turtle with a relatively large, black-coloured head with several cream, white or yellow stripes. The head pattern comprises a horizontal stripe passing from the tip of the snout, above the eye to the back of the head; this stripe is broken behind the eye. A second unbroken stripe runs from the first stripe on the top of the snout, beneath the eye to the base of the neck. The snout has six vertical stripes. The pupil has ivory rings. The moderately arched, oval carapace ranges from light to dark brown with a characteristic cream or yellow border, and lacks posterior serrations. The yellow or cream plastron has large dark brown or black blotches on each scute, and has scute rings. Two dark marks are also found on the bridge. The limbs are grey to black with a narrow yellow outer border. The tail is very short. **Behaviour:** Has a specialized diet of snails, although also eat worms, aquatic insects, crustaceans, and small fish. **Habitat:** Slow-moving waterbodies, including small streams, canals, ponds, marshes, and rice paddies. **Range:** Fairly abundant. Recorded in northeastern Thailand. **Regional Range:** Vietnam, Laos, Cambodia, and Thailand. **International Conservation Status:** IUCN: Vulnerable; CITES: Appendix II.

23

Flat-Shelled Turtle Genus *Notochelys*

This genus is tropical and aquatic. The plastron has a freely movable hinge between the pectoral and abdominal scutes. The posterior margin of the carapace is serrated. It is unique in having more than 5 scutes present, usually six or seven vertebral scutes. Worldwide: 1 species. Thailand: 1 species.

24 Flat-shelled Turtle *Notochelys platynota*

Carapace Length: Up to 32.0 cm.

A medium-sized turtle with a dorsally flattened carapace that is elongated, oval, and has a low interrupted medial keel. The carapace is coloured greenish-brown to yellowish-brown or reddish-brown. Each scute contains a dark spot or radiation. Juveniles have a bright green shell with small black dots. The plastron and bridge are yellow to orange. The head, neck, and limbs are brown. **Behaviour:** Herbivorous, feeding on aquatic plants. **Habitat:** Found in slow-moving streams, swamps, marshes, and ponds in Malayan mixed dipterocarp forest. Favours wetlands with soft bottoms and abundant vegetation. **Range:** Found in lower peninsular Thailand. Recorded in Trang, Songkhla, Pattani, and Narathiwat provinces. **Regional Range:** Vietnam, Thailand, Malaysia, Singapore, and Indonesia. **International Conservation Status:** IUCN: Vulnerable; CITES: Appendix II.

24

Pond Turtle Genus *Siebenrockiella*

This group is tropical and aquatic. It has a rigid plastron, lacking a hinge. The triturating surface of the maxilla is broad posteriorly, but medially narrowed. Worldwide: 1 species. Thailand: 1 species.

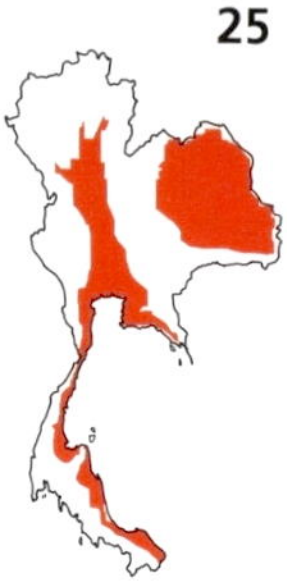

25 Black Pond Turtle *Siebenrockiella crassicollis*

Carapace Length: Up to 20.0 cm.

An almost totally dark-brown to black turtle. The carapace is oval, depressed, and strongly serrated posteriorly. The plastron and bridge are black, with some brownish streaks. The head is black to dark grey with a faded, cream or yellow spot behind each eye. Has cream to tan jaws. **Behaviour:** Elusive by nature. Carnivorous, feeding on slugs, snails, shrimps, and worms. It also eats carrion. **Habitat:** Found in slow-moving rivers, streams, marshes, swamps, lakes, and ponds. Also found in peat swamp forest. Prefers waterbodies with soft bottoms with abundant edge vegetation. Slow in movements. **Range:** Northern, northeastern, central and peninsular Thailand. Fairly common. Recorded in upper and lower central plains including Bangkok, as well as Chumphon, Surat Thani, and Narathiwat provinces. **Regional Range:** Myanmar, Vietnam, Laos, Cambodia, Thailand, Malaysia, Singapore, and Indonesia. **International Conservation Status:** IUCN: Vulnerable; CITES: Appendix II.

25

Tortoises Family Testudinidae

Members of this family are terrestrial. The limbs are more or less cylindrical; the hind feet are club-shaped. The digits are not webbed. The head is shielded above. 3 species occur in Thailand.

KEY TO THE THAI TESTUDINIDAE

1 a	Supracaudal shield single	*Indotestudo*
b	Supracaudal shield double	*Manouria*

Key modified from Taylor (1970).

Asian Elongated Tortoises Genus *Indotestudo*

This group has forelimbs which are club-like and the forefoot has five claws. The fifth, sixth, and seventh marginals touch the second pleurals. The humero-pectoral seam crosses the entoplastron. Worldwide: 2 species. Thailand: 1 species.

26 **Elongated Tortoise** *Indotestudo elongata*

Carapace Length: Up to 27.5 cm.

The domed carapace is strongly elongated, being flattened dorsally with sloping sides. The carapace is yellowish-brown or olive, with black blotches on the vertebrals and pleurals. The well-developed plastron has a deep anal notch; both the plastron and bridge are coloured yellow and unmarked. During the breeding season the skin around the eyes and nostrils become bright reddish pink. The limbs are brown to olive. **Behaviour:** Terrestrial. Herbivorous, eating fruits and flowers. Especially fond of mushrooms; congregations are often found where mushrooms flourish after heavy rains. Hide among rocks or deep in tree buttresses. **Habitat:** Favours mixed deciduous forest at higher altitudes, but also inhabits (peninsular and southeastern) monsoonal evergreen forest. **Range:** Found throughout the country, though with a scattered distribution. Recorded in Tak, Loei, Nakhon Ratchasima, Sa Kaeo, Chanthaburi, Chonburi, Uthai Thani, Kanchanaburi, Surat Thani, Krabi, Narathiwat, and Yala provinces. **Regional Range:** India, Bangladesh, Myanmar, China, Vietnam, Laos, Cambodia, Thailand, and Malaysia. **International Conservation Status:** IUCN: Endangered; CITES: Appendix II.

26

Asian Tortoises Genus *Manouria*

These tortoises have forelimbs which are club-like and the forefoot has five claws. They lack hinges on either the carapace or plastron. The supracaudal is divided into two scutes. Worldwide: 2 species. Thailand: 2 species.

KEY TO THE THAI *MANOURIA*

1 a	Carapace uniformly dark-brown, olive, or black; posterior marginals only slightly serrated; pectoral scutes may not meet at plastral midline; several very large pointed tubercles on each thigh	*M. emys*
1 b	Carapace yellow brown or brown with dark seams; posterior marginals strongly serrated; pectoral scutes always meet at plastral midline; a single, large conical tubercle surrounded by much smaller scales on each thigh	*M. impressa*

Key modified from Taylor (1970).

27 Asian Brown Tortoise *Manouria emys*

Carapace Length: Up to 60.0 cm.

The largest tortoise in the country, with an oval domed carapace with sloping sides. The carapace is uniformly olive, dark brown, or black. The posterior marginals are only slightly serrated; in the similar Impressed Tortoise these marginals are strongly serrated. Both the anterior and posterior marginals are upturned. **Behaviour:** Diurnal, foraging for succulent vegetation. Lays clutches of up to 30 eggs. **Habitat:** Found in hill evergreen forest, peninsular monsoonal evergreen forest, and Malayan mixed dipterocarp forest. **Range:** Found in northern, western, and peninsular Thailand. Recorded in Mae Hong Son, Uthai Thani, Kanchanaburi, Surat Thani, Nakhon Si Thammarat, Songkhla, Trang, and Narathiwat provinces. **Regional Range:** Myanmar, Vietnam, Laos, Cambodia, Thailand, Malaysia, Singapore (extinct), and Indonesia. **International Conservation Status:** IUCN: Endangered; CITES: Appendix II.

27a

27b

28 Impressed Tortoise *Manouria impressa*

Carapace Length: Up to 27.0 cm.

The carapace ranges from yellowish-brown to brown with dark seams. The marginals have large black blotches. The plastron is yellowish-brown with dark seams. The large head has a non-projecting snout. Unmistakable. **Behaviour:** Seems to be most active at dusk. Herbivorous, feeding on plant matter, bamboo shoots, and fallen fruits. Terrestrial. **Habitat:** Found in lower and upper montane evergreen forest. **Range:** Northern, western, northeastern (west), and peninsular Thailand. Recorded in Chiang Mai, Tak, Loei, and Narathiwat provinces. **Regional Range:** Myanmar, China, Vietnam, Laos, Cambodia, Thailand, and Malaysia. **International Conservation Status:** IUCN: Vulnerable; CITES: Appendix II.

28

Lizards and Snakes

Order Squamata

The Squamata, or the scaled reptiles, are the largest recent order of reptiles, comprising all lizards and snakes. It is the third largest order of vertebrates, with over 9,000 species. Members of the order are distinguished by their skins, which bear horny scales or shields. They also possess movable quadrate bones, making it possible to move the upper jaw relative to the neurocranium. This is particularly visible in snakes, which are able to open their mouths very wide to ingest comparatively large prey.

SUBORDER SAURIA

The order contains all the reptiles with scales, comprising all the lizards and snakes. The suborder Sauria, commonly called lizards, with their typical characteristics, also contains many species which are limbless. This group contains eight families, namely the agamid lizards (Agamidae), the cat geckos (Eublepharidae), the geckos (Gekkonidae), the Old-world lizards (Lacertidae), the glass lizards (Anguidae), the skinks (Scincidae), the snake skinks (Dibamidae), and the monitor lizards (Varanidae).

KEY TO THE FAMILIES OF THAI LIZARDS

1	**a**	Diminutive, worm-like lizards with large plate covering most of snout; eyes covered; arms absent; legs present as two small flaps at sides of vent in males only; preanal pores present, scales uniform, cycloid, imbricate	*Dibamidae*
	b	Lizards of various sizes, usually not with a single scale covering most of snout, or, if so, no sexual dimorphism in limbs	**2**
2	**a**	Very large lizards, 1-3 m in length, covered with small tuberculate juxtaposed scales, each surrounded by a row of minute tubercles, tongue long, slender, deeply bifid, retractile into a sheath; always pentadactyl; an elongate tail; head with numerous small scales above	*Varanidae*
	b	Smaller lizards, under 1 m in length; tongue not as described above; without rows of minute tubercles surrounding scales; limbs may or may not be pentadactyl	**3**
3	**a**	Scales covering top and sides of head not enlarged or symmetrical	**4**

b Scales covering top and sides of head enlarged, symmetrical, or if with small scales and tubercles **5**

4 a Vertebrae procoelian; parietal bone single; eye with thick movable connivent eyelids *Eublepharidae*

b Vertebrae amphicoelian; parietal bone divided; eyelids non-movable *Gekkonidae*

5 a Teeth acrodont, varying in shape and not replaced; tongue thick; femoral pores absent (except in the genera *Liolepis, Physignathus* and *Mantheyus*) *Agamidae*

b Not as above **6**

6 a Legless *Anguidae*

b Two pairs of legs, if not, ear-opening present **7**

7 a Inguinal or femoral pores; body covered with large rhomboidal keeled shields forming longitudinal series; tail extremely long (Thai form) *Lacertidae*

b No femoral, preanal or inguinal pores; body covered with cycloid imbricating scales above and below; tail usually about equal to snout to vent length, rarely twice as long *Scincidae*

Family **Eublepharidae**

Members of this family have movable eyelids and vertical pupils. Together with members of the Family Gekkonidae, they are some of the very few lizards with a voice mechanism. This group is sometimes placed in a subfamily Eublepharinae of the Family Gekkonidae. Only one genus, *Aeluroscalabotes*, represented by one species, is known from Thailand.

29 Cat Gecko *Aeluroscalobates felinus*

Snout to Vent: 11.0 cm. Tail: 7.0 cm.

This strange-looking lizard has both upper and lower eyelids, a unique characteristic, and is capable of closing its eyes. It has a disproportionately large head, more than twice as wide as the neck. The male has a uniformly brown-coloured body and tail which are strongly compressed. The female is more variegated with paler patterns on the body and dark spots on the tail. The lip, chin, throat, and venter are cream-coloured. Males have preanal pores. The tail is relatively short. **Behaviour:** Nocturnal, foraging by slowly crawling in vegetation overhanging waterways. Diet consists of arthropods. **Habitat:** Found in Malayan mixed dipterocarp forest and peat swamp forest. **Range:** Recorded in Narathiwat, Satun, and Yala provinces in extreme southern Thailand. **Regional Range:** Thailand, Peninsular Malaysia, Singapore, and Indonesia (Kalimantan).

29

Geckos Family Gekkonidae

Members of this family have soft granules or scales often intermixed with larger tubercles covering the dorsal and lateral parts of the body. Many members of the genera have specialized structures on the digits that enable them to run or crawl about a ceiling or wall as easily as on a floor. These consist of widened overlapping lamellae which have fine hair-like processes, the purpose being to cause adhesion to a smooth surface; among Thai species, these are evident only in arboreal species. The claws are not lost. The tails of species in this family are fragile and are easily lost or broken, and soon regenerate. Nine genera of the Gekkonidae are known from Thailand. These comprise the Bent-toed Geckos *Cyrtodactylus* (16 species), the Rock Geckos *Cnemaspis* (7 species), the Ground Geckos *Dixonius* (3 species), the Four-clawed Geckos *Gehyra* (4 species), the Dwarf Geckos *Hemiphyllodactylus* (2 species), the Tokay Geckos *Gekko* (5 species), the House Geckos *Hemidactylus* (4 species), and the Parachute Geckos *Ptychozoon* (4 species).

KEY TO THE THAI GEKKONIDAE

1	**a**	Digits slender, lacking enlarged, adhesive lamellae	**2**
	b	Digits at least partially widened with at least some enlarged adhesive lamellae	**3**
2	**a**	Pupil round	*Cnemaspis*
	b	Pupil a vertical slit; or when further closed appears as a vertical series of small openings	*Cyrtodactylus*
3	**a**	A single terminal pair of widened lamellae between which the claw is retractile.	*Dixonius*
	b	More than a single pair of adhesive lamellae	**4**
4	**a**	Tail strongly lobulate laterally, sometimes widened at tip; a lateral wing-like expansion on each side of body and head	*Ptychozoon*
	b	Tail not lobulate, a lateral fringe of skin present on side of body or not	**5**
5	**a**	A fringe of skin from axilla to groin; tail with a denticulate fringe	*Hemidactylus-part*
	b	No skin fringe along side of body	**6**
6	**a**	Subdigital lamellae undivided; terminal joints of outer digits united with expanded part of digit; inner digit clawless	*Gekko*

b	Terminal joints of digits not united with the widened lamellae, subdigital lamellae divided	**7**
7 a	Inner digit well-developed	**8**
b	Inner digit vestigial, without free terminal joint, clawless or with a minute claw; subdigital lamellae divided or single	*Hemiphyllodactylus*
8 a	Inner digit well-developed, without free terminal joint, the claw minute, often concealed.	*Gehyra*
b	Inner digit well-developed, with a free terminal joint, clawless terminal joint	*Hemidactylus-part*

Key modified from Taylor (1963).

Bent-Toed Geckos Genus *Cyrtodactylus*

Members of this genus are largish species, variously coloured and conspicuously marked. The dorsal surface is composed of small granular scales intermixed with larger keeled tubercles. The digits are slender with claws, the tail is cylindrical and longer than the head and the body. Preanal and/or femoral pores are present. The pupil is vertical. Sixteen species have been recorded in Thailand.

KEY TO THE THAI *CYRTODACTYLUS*

1 a	Foot partially webbed; tail with lateral denticulations	*2*
b	No web present on hand or foot	**3**
2 a	Nine preanal pores on either side; brown with W-shaped dorsal marks in young which may disappear in adults	*C. brevipalmatus*
b	Seven preanal pores on either side, 8-9 femoral pores	*C. interdigitalis*
3 a	Dorsal pattern has four parallel longitudinal stripes	*C. quadrivirgatus*
b	Dorsal pattern has crossbands, spots, or blotches	**4**
4 a	Dorsal pattern has pale or dark crossbands	**5**
b	Dorsal pattern has dark spots or blotches	**11**
5 a	Preanal groove absent	**6**
b	Only preanal pores present, numbering eight, in two parallel rows lying in a deep groove but contiguous, at right angles, with a long series of femoral pores; a curved, light-edged band from eye around back of occiput; body with transverse, light-edged, dark bands	*C. pulchellus*
6 a	Preanal and femoral pores absent	*C. thirakupti*
b	Preanal present or both of preanal and femoral pores present	**7**
7 a	Male with preanal pores only, no femoral pores	*C. sumonthai*
b	Preanal and femoral pores present	**8**

8 a	Mid-body scales number more than 50 (58-65)	*C. consobrinus*
b	Mid-body scales 50 or fewer	**9**
9 a	Dorsal pattern comprises dark crossbars; does not inhabit caves	*C. intermedius*
b	Dorsal pattern is pale bars, inhabits caves	**10**
10a	Mid-body scales number 38, femoral pores continuous to preanal pores (32 in total)	*C. chanhomeae*
b	Mid-body scales number 34; femeral pores separated from preanal pores (21 in total)	*C. tigroides*
11a	Male with preanal pores only, no femoral pores	**12**
b	Preanal and femoral pores present	**15**
12a	Top of head with pale or dark spots	**13**
b	Top of head without dark spots; mid-body scales 34-40	*C. angularis*
13a	Top of head with pale spots; post occipital band with curved line at the front and back edges	**14**
b	Top of head without dark spots; post occipital band with lobule-shaped line at the front edge	*C. papilionoides*
14a	Male with 4 preanal pores; mid-body scales fewer than 34-38	*C. oldhami*
b	Male with 7-8 preanal pores; mid-body scales 32	*C. peguensis*
15a	Mid-body scales fewer than 30 (22)	*C. variegatus*
b	Mid-body scales more than 30 (32-38)	*C. jarujini*

Key modified from Taylor (1963); Das (1997).

30 Web-footed Bent-toed Gecko *Cyrtodactylus brevipalmatus*

Snout to Vent: 7.2 cm. Tail: 7.7 cm.

A smallish, medium-sized gecko with partially webbed feet—a feature that distinguishes this species from all other members of the genus. The body is pale greyish-brown with numerous small keeled tubercles, which sometimes have indistinct darker spots edged with dark brown. The forehead is chocolate-brown, with white markings. The underside is creamy white. On either side of the body are nine very large preanal pores. There is no pubic groove. The tail has lateral denticulations. Young animals have W-shaped dorsal marks; these disappear in the adults. **Behaviour:** Nocturnal. It forages on the ground and on tree trunks, as well as human dwellings in the forest interior. During the daytime, known to hide under dead bark on tree trunks. It has a characteristic habit of coiling its tail. **Habitat:** Found in dry evergreen forest and peninsular monsoonal evergreen forest up to 1,000 m. **Range:** Found in western and peninsular Thailand. Recorded in Tak Province; Huai Kha Khaeng in Uthai Thani Province; Kaeng Krachan in Petchaburi Province; and Khao Luang in Nakhon Si Thammarat Province. Generally regarded as rare. **Regional Range:** Endemic to Thailand.

30

31 Ulber's Bent-toed Gecko *Cyrtodactylus interdigitalis*

Snout to Vent: 8.0 cm. Tail: 8.9 cm.

A medium-sized *Cyrtodactylus* gecko, with distinct, flat and very finely scaled skin webbing between the second and third, and the third and fourth hind toes and likewise between all the toes of the forefeet; this feature is diagnostic for this species. The body is rich brown on the dorsal surface, with six transverse somewhat darker bands. It has 18-22 irregular, long rows of smooth or weakly keeled tubercles. The sides are lighter brown. The venter is yellow brown. The tail is short, and ringed in light and darker brown, with dark-brown flecks on the tail base. The males have 14 preanal pores; in the Web-footed Bent-toed Gecko the male only has nine preanal pores. **Behaviour:** Nocturnal. Lives in caves or in adjacent fig trees with numerous crevices close to caves. At dusk, starts foraging on the ground or among boulders in close proximity to caves. Known to coil its tail like the Web-footed Bent-toed Gecko. **Habitat:** Inhabits caves in the vicinity of limestone hills. Also occurs in evergreen forest close to caves. **Range:** Found in western parts of northeastern Thailand. Recorded in Phu Luang Wildlife Sanctuary, Loei Province and Nam Nao National Park, Petchabun Province. **Regional Range:** Laos and Thailand.

31

32 Peters' Bent-toed Gecko *Cyrtodactylus consobrinus*

Snout to Vent: 12.0 cm. Tail: 15.7 cm.

Adults are much darker than young, being nearly black with pale narrow white or yellow bars on the body and limbs. The dorsal surface has 18-20 tubercles in irregular rows. Young have finely speckled brown body with irregular, black-edged, cream lines. Supralabials are black and grey spotted with white. All ventral surfaces are pale grey, except tail; the sub-caudals are dark with narrow pale bands. **Behaviour:** Nocturnal. Found on tree trunks, particularly of large dipterocarps. Live in loose family groups of 2-4 animals. Adults forage high up tree trunks, while young forage close to the ground. Upon being disturbed, they tend to move higher up the trees, and hide in crevices. **Habitat:** Inhabits Malayan mixed dipterocarp forest in primary and secondary forest in lowlands. **Range:** Recorded in Narathiwat Province; Hala Bala, Bang Lang and Betong, Yala Province and Satun Province in extreme southern Thailand. **Regional Range:** Thailand, Peninsular and Eastern Malaysia, and Indonesia (Sumatra).

32

33 Intermediate Banded Bent-toed Gecko *Cyrtodactylus intermedius*

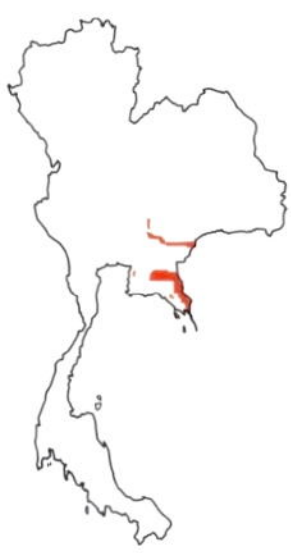

Snout to Vent: 6.8 cm. Tail: 7.7 cm.

A small gecko, superficially similar to Banded Bent-toed Gecko. The body is light brown, marked with four distinct bands of dark brown or blackish-brown each edged with yellow or cream. The top of the head is nearly uniform brownish; the head lacks the spotted markings of Tiger Bent-toed Gecko. Another cream-edged, dark-brown band runs from the eye around the occiput to the other eye. The tail is banded. The venter is creamy yellow. It has 40-50 rows of scales around mid-body. It has 8-10 preanal pores. **Behaviour:** Nocturnal. Forages closes to streams, particularly dried-up creeks. During the daytime, hides under bark and in rotten branches on the ground. Young forage on vegetation close to streams. **Habitat:** Inhabits southeastern monsoonal evergreen forest. **Range:** Found in southeastern Thailand as far northward as the Dong Phaya Yen Forest Complex, in Nakhon Ratchasima Province. **Regional Range:** Cambodia and Thailand.

33

34 Jarujin's Bent-toed Gecko *Cyrtodactylus jarujini*

Snout to Vent: 9.0 cm. Tail: 11.6 cm.

Pale brownish-bodied gecko with a series of dark spots arranged as irregular bands running down the back; there are several dark spots scattered on the top of the head. The tail is also brownish with black or dark-brown bands extending along its length. A dark-brown band passes from the back of the eye to above the ear opening, There is no broad band around the base of the neck. It has a series of somewhat specialized preano-femoral scales. The male has 52-54 pores preano-femoral pores in an angular series of scales. There are 32-38 scale rows between the ventrolateral folds. **Behaviour:** Strictly nocturnal, being active quite late at night. Forages on large sandstone boulders, especially on the vertical faces. On being disturbed, it quickly crawls to hide in overhanging vegetation. During daylight hours, spends its time in large crevices between the sandstone boulders. **Habitat:** Occurs in dry dipterocarp forest with plenty of bamboo, in sandstone hills. **Range:** Recorded in Phu Wua Wildlife Sanctuary, Bueng Kan Province in northeast Thailand. **Regional Range:** Laos and Thailand.

34

35 Angular-spotted Bent-toed Gecko *Cyrtodactylus angularis*

Snout to Vent: 9.2 cm. Tail: 9.5 cm.

Greyish-bodied gecko with four characteristic butterfly-shaped or W-shaped black or brown marks running down the back; between these markings are numerous black flecks. The tail is also greyish with black or brown bands extending along its length. A light-brown, black-edged band passes from the snout, through the eye to the back of occiput, where it forms a broad band around the base of the neck. It has a series of somewhat specialized femoral scales. The male has six preanal pores in an angular series of scales, sometimes not strongly developed. There are 40 scale rows between the ventrolateral folds. **Behaviour:** Nocturnal. Forages on the forest floor late at night, crawling slowly among the leaf-litter. Sometimes found on forest roads. When disturbed, it can remain motionless for up to 30 minutes. Young tend to forage in low vegetation, off the ground. Spends the daytime under loose bark at the base of tree trunks, or in rotting logs on the ground. Voice is a loud same-scale knocking sound (klock-klock-klock) 5-35 times every half second. **Habitat:** Found in dry evergreen forest, as well as edges of dry dipterocarp forest. **Range:** Restricted to the mountain range of central Thailand, being recorded in Saraburi Province and Nakhon Ratchasima Province in the eastern central plains. **Regional Range:** Endemic to Thailand.

35

36 Butterfly Bent-toed Gecko *Cyrtodactylus papilionoides*

Snout to Vent: 8.1 cm. Tail: 9.5 cm.

Dorsal ground coloration ranges from beige-brown to strong middle-brown, with butterfly wing to banded markings; the dorsum pattern is irregular blotches. The arrangement of 12-14 long rows of dorsal tubercles is orderly. The forward edge of the neckband is generally arranged in a wave pattern. The venter is dirty white. Very similar to Angular Spotted Bent-toed Gecko, but characterised by an unbroken row of 29-33 enlarged preanofemoral scales, in which the middle scales possess 4-6 distinct preanal pores, homologous granular gular scales without scattered tubercles, and 30-40 rows of ventral scales over the center of the abdomen. The number of caudal tubercles is 6-8 beginning from the third whirl of the tail base. Lacks distinctly enlarged scales beneath the tail. **Behaviour:** Nocturnal. Found on the ground, but also found in small cliffs and rock faces. Also found in limestone caves, where it climbs on the walls and hides in cracks and crevices. When seen in the daytime, tends to move upward on the walls and toward darker parts of the caves. Often forages around the cave entrance. Voice is a faint clicking sound. **Habitat:** Found in limestone caves, in dry evergreen forest. **Range:** Recorded in Chaiyaphum, Petchabun, and Loei provinces in western northeast Thailand. **Regional Range:** Endemic to Thailand.

36

37 Banded Bent-toed Gecko *Cyrtodactylus pulchellus*

Snout to Vent: 11.5 cm. Tail: 14.4 cm.

A large gecko, with four transverse, cream-edged, dark bands on the olive green, background, body colour. The tail is also banded, with at least seven dark-brown or blackish bands. The head is uniformly tan, with a curved or angular, light-edged dark brown band running from the eye around the back of the occiput: this head marking is absent in the similar Four-lined Bent-toed Gecko. The chin and venter are creamy white. The subcaudal coloration is lavender to violet. **Behaviour:** Nocturnal. Climbs on tree trunks and hides in rock crevices. **Habitat:** Occurs in peninsular monsoonal evergreen forest and Malayan mixed dipterocarp forest to high altitudes. **Range:** Peninsular Thailand, from Nakhon Si Thammarat Province southward. **Regional Range:** Myanmar, Thailand, Peninsular Malaysia, and Singapore.

37

38 Four-lined Bent-toed Gecko *Cyrtodactylus quadrivirgatus*

Snout to Vent: 6.5 cm. Tail: 7.0 cm.

Smallish, greyish-brown gecko with three or four black, longitudinal stripes running from the eye down the length of the body. The tail has transverse black bands separated by grey bands. Some individuals have distinctive lavender or violet limbs, being more intensely coloured on the digits. The head is tan-coloured. It lacks the dark band passing from the eyes to the back of the occiput, as found in other geckos. The chin and venter are uniformly white with black speckling. The male has four preanal pores. **Behaviour:** Nocturnal. Found close to the ground, up to 2 m. Lives in trunks of dead trees, fissures in logs and in spiny palms, vines, saplings and other vegetation. **Habitat:** Inhabits peninsular monsoonal evergreen forest, Malayan mixed dipterocarp forest, and peat swamp forest. **Range:** Peninsular Thailand. Recorded in Trang Province, Pa Phru Wildlife Sanctuary; Narathiwat Province, and Hala Bala Wildlife Sanctuary, Yala Province. **Regional Range:** Thailand and Peninsular Malaysia.

38

39 Oldham's Bent-toed Gecko *Cyrtodactylus oldhami*

Snout to Vent: 6.8 cm. Tail: 7.0 cm.

A smallish gecko, with a brown dorsum marked by a series of elongated or rounded cream spots arranged in 4 broken, longitudinal lines running down the back. In the northern part of the range, these may be solid brown stripes from the eyes to the mid-body. Top of head uniformly brown, which.contrasts with a cream-edged, dark brown band running from eye to eye behind the occiput. The tail is dark brown with dorsal cream bars.

The chin and venter are dirty white. The male has one or two preanal pores on each side (occasionally absent). **Behaviour:** Nocturnal. Hunts on the ground and on earthern banks, very rarely climbing tree trunks. **Habitat:** Found in dry evergreen forest and peninsular monsoonal evergreen forest. **Range:** Found in western and peninsular Thailand. Recorded in Umphang Wildlife Sanctuary, Tak Province and other localities in the Western Forest Complex, as well as many sites in peninsular Thailand. **Regional Range:** Myanmar and Thailand.

39

---

40 Spotted Bent-toed Gecko *Cyrtodactylus peguensis*

Snout to Vent: 6.7 cm. Tail: 6.1 cm.

Moderately flattened head has a creamy white reticulum (a sharply defined stripe from behind the eye around the occiput) enclosing about 18 dark-brown or black spots. The back is also marked by a median series of dark brown or black spots, as well as a lateral series of transversely widened dark spots all in a reticulum of cream-white. The tail has nine black rings, separated by narrower white rings. **Behaviour:** Nocturnal. Found mainly on the ground, near the base of trees and on large stones. Young stay in low vegetation. **Habitat:** Inhabits dry evergreen and peninsular monsoonal evergreen forest. **Range:** Found in western and peninsular Thailand. **Regional Range:** Myanmar and Thailand.

40

---

41 Variegated Bent-toed Gecko *Cyrtodactylus variegatus*

Snout to Vent: 7.0 cm. Tail: 10.0 cm.

Grey above, beautifully spotted and marbled with black, set off with subdued white. On the back the markings appear as irregular bands, paler internally, and blackish on their zigzag borders. The head is spotted above with a broad dark streak with whitish border behind each eye and continuing irregularly around the occiput. Lower parts are whitish, freckled on the tail with black, the terminal third being almost totally blackish. The tail is irregularly banded above. The male has a continuous series of 32 preanal and femoral pores. **Behaviour:** Nocturnal. Lives in caves. Searches for food near the cave entrance and among nearby boulders. **Habitat:** Inhabits caves in limestone areas. **Range:** Found in western Thailand. **Regional Range:** Myanmar and Thailand.

41

42 Sumontha's Cave Gecko *Cyrtodactylus sumonthai*

Snout to Vent: 7.1 cm. Tail: 9.2 cm.

Smallish, slender-bodied gecko with a dorsal pattern comprising alternating whitish to greenish bands and brownish bands. There is one brown band across the nape; two across the trunk and one across the sacrum. Long, slender limbs and digits. Distinguished from all congeners by slender body, large homogenous body scalation, having only 12 rows of small tubercles, elongate digits and tail; two precloacal pores, and an absence of a precloacal groove and femoral pores. **Behaviour:** Cave-dwelling. Climbs high on the wall of the caves. Slow-moving. **Habitat:** Inhabits limestone caves. **Range:** Endemic to caves of southeastern Thailand. Range restricted to Khao Wong, in Rayong Province. **Regional Range:** Endemic to Thailand.

42

43 Thirakhupt's Cave Gecko *Cyrtodactylus thirakhupti*

Snout to Vent: 8.0 cm. Tail: 9.2 cm.

Small, slender-bodied gecko with elongate, slender limbs and digits. Dorsal pattern comprises asymmetrical pale yellowish bands with very dark brown borders, on a lighter brown, basal body coloration. A greyish stripe, bordered above with dark brown, passes from the eye to the first transverse band. The tail has bright white annuli posteriorly. It has only 14 rows of small tubercles and an absence of a precloacal groove. **Behaviour:** Cave-dwelling. **Habitat:** Inhabits limestone caves. **Range:** Endemic to caves in Surat Thani Province, peninsular Thailand. **Regional Range:** Endemic to Thailand.

43

44 Chanhome's Cave Gecko *Cyrtodactylus chanhomeae*

Snout to Vent: 7.9 cm. Tail: 8.0 cm.

A slender-bodied gecko. Dorsal pattern comprises alternating purplish-brown dark bands with creamish-yellow bands (five paler bands between the limb insertions). Dorsal scalation has 16-18 rows of keeled tubercles. The venter is pale brown. Long limbs and digits. The iris is greenish brown. The male has a continuous series of 32-34 pore-bearing, precloacal-femoral scales. **Behaviour:** Cave-dwelling. **Habitat:** Inhabits limestone caves. **Range:** Found in the limestone hills (caves) in the Central Plains. Range restricted to Saraburi Province. **Regional Range:** Endemic to Thailand.

44

45 Tiger Cave Gecko *Cyrtodactylus tigroides*

Snout to Vent: 8.3 cm. Tail: 11.5 cm.

Like other congeneric cave-dwelling geckos, it is a small, slender-bodied gecko with long, slender limbs and digits. The base colour is mid-brown, with yellowish-cream bands with well-defined, dark brown borders. The dorsum of the head is yellowish-cream with well-defined, symmetrical, mid-brown markings with dark brown borders. It has 13 rows of small keeled tubercles. **Behaviour:** Cave-dwelling. Forages quite far from the cave entrances for prey. Found hunting from mid-evening onward. In the daylight hours, found deep in caves or crevices. **Habitat:** Inhabits limestone caves. **Range:** Endemic to caves of Sai Yok National Park, Kanchanaburi Province, western Thailand. **Regional Range:** Endemic to Thailand.

45

Rock Geckos Genus *Cnemaspis*

Members of this genus are diminutive and have depressed bodies, cylindrical tails, and slender, clawed digits. The body is either granular or has tubercles mixed with the granules. The pupil is round. An "eyelid" is distinct all around the eye. Males may have preanal or femoral pores, or both. Seven species are found in Thailand.

KEY TO THE THAI *CNEMASPIS*

1 a	Males with femoral and preanal pores	*C. kandiana*
b	Males lack femoral pores or lack of both preanal and femoral pores	**2**
2 a	Preanal and femoral pores are absent in male. Proximal subdigital scales much larger than distal scales	*C. phuketensis*
b	Preanal pores in males present. Proximal sub-digitals not, or little, larger than distal sub-digitals	**3**
3 a	Ventral abdominal scales smooth, sub-caudal smooth	**4**
b	Ventral abdominal scales keeled	**5**
4 a	Snout relatively short and deep; distal sub-digitals 10-12: 16-17: 17-19: 18-20; about nine scales below the first metatarsal; adult males 34-42 mm in length, with eight preanal pores; about 24-30 scales around mid-tibia	*C. chanthaburiensis*
b	Snout relatively long and flattened; distal sub-digitals 13-17: 20-22: 21-25 21-24: 20-26; it has 12-17 scales below first metatarsal; adult males 36-51 mm in length, with 8 preanal pores; about 24-32 scales around mid-tibia	*C. kumpoli*
5 a	A yellow line along back present	*C. flavolineata*
b	No yellow line along back	**6**
6 a	Median subcaudal scales keeled, not distinctly larger than adjacent scales usually two tubercles at the side of the vent; usually a dark blotch or ocellus above arm insertion; no median line on the throat; males with 2-6 preanal pores; 27.7-46.7 mm in length.	*C. narathiwatensis*
b	Median subcaudal scales in a series of keeled, weakly pointed, enlarged scales; males with 4-8 preanal pores; no dark blotch above arm insertion	*C. siamensis*

Key modified from Taylor (1963); Dring (1979).

46 Kandy Rock Gecko *Cnemaspis kandiana*

Snout to Vent: 3.0 cm. Tail: 3.8 cm.

Diminutive brown species with numerous lighter-brown spots, particularly evident on the sides of the neck and on digits. Ventral surface is much lighter. Has 2-3 preanal pores which are widely separated from 3-5 femoral pores. Has some small spine-like scales scattered on flanks. **Behaviour:** Diurnal. Found under stones and logs. **Habitat:** Found in peninsular monsoonal evergreen forest, in hilly country at low altitudes. **Range:** A restricted range in northern peninsular Thailand. Recorded in Tasan and Namtok Kra Po, Chumphon Province; and Koh Yan Yan and Koh Lang, Ranong Province. **Regional Range:** Sri Lanka and Thailand.

46

47 Siamese Rock Gecko *Cnemaspis siamensis*

Snout to Vent: 3.9 cm. Tail: 5.0 cm.

Very small. Entire body is mottled brown and light brown. Above grey-green with darker-brown markings, which appear to form transverse bars on the back. The body has numerous rows of enlarged conical, trihedral, or tricarinate scales. Has a pair of lines behind the eyes which converge to a light spot behind the nape. The dorsal surface of the slightly compressed tail has 10 bars. The venter is whitish-yellow. Very slender, clawed digits. Male has preanal pores. **Behaviour:** Diurnal. Primarily a rock dweller but also occurs on tree trunks and buttress roots. **Habitat:** Inhabits peninsular monsoonal evergreen forest. **Range:** Found in peninsular Thailand from Ratchaburi Province southward. **Regional Range:** Endemic to Thailand.

47

48 Chanthaburi Rock Gecko *Cnemaspis chanthaburiensis*

Snout to Vent: 4.1 cm. Tail: 4.9 cm.

Very small. Chocolate-brown dorsal surface has characteristic white markings running down either side of its spine, from the nape to the base of the tail. Continuation of white markings on the brown tail form narrow bands.

A ridge of tubercles borders the anterior margins of the ears and another from the ear to the nape. The belly is mottled light brown. Limbs light to medium chocolate-brown. The snout is relatively short. Male has 7-9 pre-anal pores; female has none. Ventral abdominal and sub-caudal scales are smooth. **Behaviour:** Diurnal. Found under stones and logs. **Habitat:** Occurs in southeastern monsoonal evergreen forest. **Range:** Found in southeastern Thailand. Recorded in Khao Soi Dao Wildlife Sanctuary, Khao Sabab, and Khao Kitchakut National Park in Chanthaburi Province; Khao Chamao-Khao Wong National Park in Rayong Province; and Khao Ang Ru Nai Wildlife Sanctuary in Chachaengsao Province. **Regional Range:** Laos, Cambodia, and Thailand.

48

49 Malayan Rock Gecko *Cnemaspis narathiwatensis*

Snout to Vent: 4.8 cm. Tail: 4.5 cm.

A small rock gecko, with head and body less than 5 cm in length. Above mottled yellowish-brown and dark brown, with five characteristic transverse bands; the two anterior ones are very bright. It usually has a dark blotch or ocellus above the arm insertion. It lacks enlarged tubercles on the tail found in conspecifics. Fingers are relatively short. Male has 2-6 pre-anal pores. **Behaviour:** Diurnal and nocturnal. Found climbing on walls of caves. **Habitat:** Found in caves in Malayan mixed dipterocarp forest. **Range:** A single record from Hala Bala Wildlife Sanctuary, Yala Province in extreme southern Thailand. **Regional Range:** Thailand and Peninsular Malaysia.

49

50 Striped Rock Gecko *Cnemaspis flavolineata*

Snout to Vent: 9.8 cm. Tail: 11.4 cm.

This rock gecko has a characteristic broad yellow band down the back, extending from the base of the neck and increasing in width near the abdomen, and then narrowing along the tail. On either side of this band, the body colour is dark mottled brown. On the sides are four narrow vertical yellowish bands, which are broken up. The digits are irregularly marked with black and yellow. The underparts are grey. **Behaviour:** Presently unknown. **Habitat:** Found in Malayan mixed dipterocarp forest. **Range:** A single specimen in Hala Bala Wildlife Sanctuary, Yala Province in extreme southern Thailand. **Regional Range:** Thailand and Peninsular Malaysia.

50

51 Kumpol's Rock Gecko *Cnemaspis kumpoli*

Snout to Vent: 5.2 cm. Tail: 6.8 cm.

Body and tail marked with blackish-grey and light-grey transverse bands. It has a horseshoe-shaped mark bordered by a lighter olive-grey line running from one eye, round the occipital to the other eye. The chin and venter are greyish-white. The subcaudal area is smooth with a median series of somewhat rounded scales (not pointed), three to each segment, the third of each group the largest. Two median rows of enlarged dorsal tubercules form straight lines. It has 3-4 minute preanal pores separated mesially by two scales. It has long slender limbs and long digits. **Behaviour:** Nocturnal. Found on tree trunks as well as under rocks on stream banks. **Habitat:** Occurs in peninsular monsoonal evergreen forest and Malayan mixed dipterocarp forest. **Range:** Found throughout peninsular Thailand from Chumphon Province southward. **Regional Range:** Thailand, Peninsular Malaysia, and Singapore.

51

52 Phuket Rock Gecko *Cnemaspis phuketensis*

Snout to Vent: 2.9 cm. Tail: 3.4 cm.

A small species of Rock Gecko. The dorsum is olive, with sinuous dark greyish markings on the nape, axilla, and inguinal regions. A dark greyish canthal stripe passing from the snout through the eye to the axilla. It has scattered, chamois-coloured, spinose tubercles on the dorsum. The limbs and segmented tail are olive with dark bands. The venter is unpatterned cream, with brown mottling on the underside of the tail. Adult males lack preanal pores. It has two semi-circular supranasals separated by a single scale, and three postnasals bordering the nasal. Lamellae under toe IV number 16-17. **Behaviour:** Diurnal. Found on the ground, in low foliage or on tree trunks. **Habitat:** Found in proximity to waterfalls in peninsular monsoonal evergreen forest. **Range:** Only recorded in Phuket, peninsular Thailand. **Regional Range:** Endemic to Thailand.

52

Ground Geckos Genus *Dixonius*

Members of this genus are completely ground-dwelling and have sturdy bodies and strong well-developed limbs for movements in the leaf litter and loose soil. Three species occur in Thailand.

KEY TO THE THAI *DIXONIUS*

1 a Body above with numerous rather large black spots; tail banded or at least when it is young **2**

b Dorsal spots absent or minute if evident, arranged in four longitudinal lines; a black stripe runs from tip of snout to tail bordered above by a cream-white stripe to shoulder; labial whitish *D. melanostictus*

2 a Body size smaller (snout to vent length 42 mm); a greater number of midbody scales (12-14 rows of tubercles and 22-26 ventrals); eight preanal pores; tail orange in adult *D. hangseesom*

b Body size larger (svl 50 mm); a lesser number of mid-body scales (10-11 rows of tubercles and 22 ventrals); six preanal pores; tail banded dark and light. *D. siamensis*

Key modified from Taylor (1963).

53 **Siamese Ground Gecko** *Dixonius siamensis*

Snout to Vent: 5.2 cm. Tail: 5.1 cm.

A small species with lips strongly barred with black and cream. The head, back, and tail is lavender brown to greyish, strongly marked with large irregular black spots. Has an indistinct series of 17 buff, pinkish, or yellow dots running from the nape along the dorsolateral length of the body to the tip of the tail. Lacks a black stripe running from tip of snout along the entire length of the body and down the tail, distinguishing the species from Black-sided Ground Gecko. Venter is yellowish. Limbs are well-developed, with a widened pair of terminal lamellae. Usually has six preanal pores. **Behaviour:** Nocturnal. Terrestrial, active on the forest floor, rocks, and tree trunks. **Habitat:** Found in lowlands in a wide range of forest habitats. **Range:** Found throughout the country. **Regional Range:** Myanmar, Vietnam, and Thailand.

53

54 Black-sided Ground Gecko *Dixonius melanostictus*

Snout to Vent: 4.2 cm. Tail: 5.4 cm.

Separated from Siamese Ground Gecko by having a black stripe from the tip of the snout running through the eye to the shoulder, where it widens and continues along the side of the body and down the tail. It has a largely patternless dorsum, with dorsal spots absent or minute if present. Nine preanal pores. **Behaviour:** Nocturnal. Found among fallen rocks at base of limestone hills. Also found in caves, foraging at the cave entrance. **Habitat:** Found in limestone forest. **Range:** Restricted to central Thailand. Recorded in Muak Lek, Saraburi Province and Pak Chong, Nakhon Ratchasima Province. **Regional Range:** Endemic to Thailand. **International Conservation Status:** IUCN: Data Deficient.

54

55 Orange-tailed Ground Gecko *Dixonius hangseesom*

Snout to Vent: 3.9 cm. Tail: 4.9 cm.

Distinguished from Siamese Ground Gecko by its smaller size, and the presence of a bold dark stripe through the eye. The dorsal body is beige, greyish, or yellowish-tan with dark-brown markings. The contrasting tail is orange, which may bear lighter, yellowish bands with dark brown margins. The male has precloacal pores. **Behaviour:** Nocturnal. Forages on the forest floor. Found in limestone crevices in daytime. **Habitat:** Found in bamboo forest on limestone hills. **Range:** Recorded from Kanchanaburi Province in western Thailand. **Regional Range:** Endemic to Thailand.

55

House Geckos Genus *Hemidactylus*

Members of this genus are relatively small species characterized by having digits with widened lamellae, divided mesially. The digits bear clawed distal phalanges. Most species have small granular dorsal scales sometimes intermixed with large tubercular granules; some lack tubercles on the body and tail. Some have flattened bodies for climbing, and a cutaneous expansion along the side of the body. The pupils are vertical. Males have preanal and/or femoral pores. Four species are found in Thailand.

KEY TO THE THAI *HEMIDACTYLUS*

1 a Body without lateral fringe-flap from axilla to groin2

b Body with lateral fringe-flap from axilla to groin3

2 a Tail not or scarcely flattened, with whorls of erect tubercular spines on each segment; without lateral denticulations. Males as numerous as females *H. frenatus*

b Tail flattened, with flat imbricating scales, lacking erect tubercular spines; edges of tail with lateral denticulations. Males unknown in Thailand *H. garnotii*

3 a Body with a narrow lateral fringe or skin-flap from axilla to groin; no fringe on side of head or neck; tail with a narrow denticulated fringe *H. platyurus*

b Body with a broad lateral fringe from axilla to groin; a fringe on side of head; tail with a broad denticulated fringe *H. craspedotus*

Key modified from Taylor (1963).

56 Spiny-tailed House Gecko *Hemidactylus frenatus*

Snout to Vent: 5.7 cm. Tail: 6.2 cm.

This small gecko has a tail with whorls of elevated spiny tubercles on the tail. The body is brownish-grey, clouded with darker grey and covered with small granules intermixed with a few scattered conical tubercles. The ventral surface is yellowish-white or white. The lamellae are dark-grey to lavender. The tail is scarcely flattened if at all. The male has a continuous series of 14-19 preanofemoral pores on each side. **Behaviour:** Nocturnal. Feeds on small insects on walls and ceilings, as well as near electric lights. Has a harsh chirping call. **Habitat:** Found in houses as well as in trees. **Range:** Found abundantly throughout the country. One of the two most common gecko species in Thailand. **Regional Range:** East Africa, India, Sri Lanka, Bangladesh, Myanmar, China, Vietnam, Laos, Cambodia, Thailand, Peninsular and Eastern Malaysia, Singapore, Brunei, Indonesia and Philippines, East Timor, Papua New Guinea, and Australia.

56

57 Garnot's House Gecko *Hemidactylus garnotii*

Snout to Vent: 6.4 cm. Tail: 8.2 cm.

Brownish-grey above, speckled or indistinctly marbled. Small white spots are commonly present. Long pointed snout. Whitish-yellow below. Lacks erect tubercular spines on the tail, which is strongly depressed. The tail has lateral serrations. Males unknown in Thailand. **Behaviour:** Found on tree trunks, rocks, and buildings. It is a parthenogenic species; the females produce fertile eggs. **Habitat:** Found in a wide ranger of habitats. **Range:** Found throughout the country. **Regional Range:** India, Myanmar, China, Vietnam, Thailand, Peninsular and Eastern Malaysia, Singapore, Brunei, Indonesia and Philippines, East Timor, Papua New Guinea, and eastward to Tahiti.

57

58 Common Frilly Gecko *Hemidactylus platyurus*

Snout to Vent: 5.7 cm. Tail: 6.3 cm.

Distinguished from all other gecko genera by presence of a lateral skin-flap running from under the arms to the groin, as well as a narrow denticulated skin-flap on the tail. This skin-flap on both the body and tail is narrower than that found in Malayan Frilly Gecko. The Common Frilly Gecko lacks any skin-flap on the side of the head and neck, features which are present on the Malayan Frilly Gecko. It has strongly webbed digits. The body is varying shades of grey, with a dorsal pattern consisting of a series of four darker median spots divided by a light-grey line. The tail is dimly banded with narrow dark-grey and wider light-grey bands. The chin and venter are creamy yellow. **Behaviour:** Nocturnal. Climbs up walls and across ceilings. **Habitat:** Found in human dwellings. **Range:** Found throughout the country. Probably the most common house-gecko in Thailand. **Regional Range:** India, Nepal, Sri Lanka, Bangladesh, Myanmar, China, Vietnam, Laos, Cambodia, Thailand, Peninsular and Eastern Malaysia, Singapore, Brunei, Indonesia and Philippines, East Timor, Papua New Guinea, and Australia.

58

59 Malayan Frilly Gecko *Hemidactylus craspedotus*

Snout to Vent: 6.2 cm. Tail: 6.5 cm.

The body and limbs are flattened, with the head scarcely distinct from the body. It also has a broad lateral skin-flap running from under the arms to the groin, as well as a broad denticulated skin-flap on the tail which distinguishes this species from all other geckos, including the Common Frilly Gecko. This skin-flap also extends on the side of the head and neck, a feature which is absent on the Common Frilly Gecko. Above generally grey, speckled with paired quadrangular brown spots. A dark line runs from the eye back to elbow. There are one or two dark longitudinal lines along the sides of the body. The venter is uniformly whitish-yellow. The tail is banded posteriorly **Behaviour:** Nocturnal. **Habitat:** Peninsular monsoonal evergreen forest and Malayan mixed dipterocarp forest. **Range:** Occurs in peninsular Thailand from Surat Thani Province southward. **Regional Range:** Thailand, Peninsular and Eastern Malaysia, and Singapore.

59

Four-Clawed Geckos Genus *Gehyra*

Members of this genus have free digits with the distal phalanges of four outer digits on hands and feet characteristically clawed. The dorsal scales are small and granular. The pupil is vertical. Males have preanal and femoral pores, usually occurring in a continuous series. Four species are found in Thailand.

KEY TO THE THAI *GEHYRA*

1 a	Sub-caudal region with a median series of largely transversely widened scales; dorsal body scales not distinctly cone-like; pores variable	**2**
b	Sub-caudal region with several series of small scales not transversely widened; 10 preanal pores on each side forming a continuous series, angular mesially, extending slightly on to femora	*G. lacerata*
2 a	Tail much widened behind vent	**3**
b	Maximum width of tail not or but slightly wider than base, scales on dorsum small, flat, cycloid, imbricating, with a slight lateral caudal fringe, not or scarcely denticulate; 15-18 femoral and preanal pores continuous, angular mesially	*G. angusticaudata*

3 a Femoropreanal pores lunate or oval, 15-20 on each side forming a doubly curved, continuous series, angular mesially, extending nearly entire length of femora; a pair of rather large supranasals in contact mesially; tail denticulate on sides *G. mutilata*

b Femoropreanal pores 11 on each side, small, subcircular forming an angle mesially and reaching about halfway along femora; supranasal smaller, separated mesially; a slight caudal fringe, somewhat denticulate *G. fehlmanni*

Key modified from Taylor (1963).

60 Western Four-clawed Gecko *Gehyra lacerate*

Snout to Vent: 5.5 cm. Tail: 3.8 cm.

A 4-clawed gecko. Dorsal surface and sides are grey. There are about 20 darker-grey spots on the head and scattered, larger grey spots on the back. Venter is cream to whitish-yellow. Tail is noticeably shorter than body. It has 10 preanal pores on each side forming a continuous series. **Behaviour:** Found under rocks or logs. **Habitat:** Inhabits forest. **Range:** Found in northeastern, southwestern, and southeastern Thailand. Recorded from Mukdahan, Kanchanaburi, Nakhon Ratchasima, Sa Kaeo, and Chonburi provinces. **Regional Range:** Endemic to Thailand.

60

61 Slender-tailed Four-clawed Gecko *Gehyra angusticaudata*

Snout to Vent: 5.7 cm. Tail: 6.1 cm.

A 4-clawed gecko with a tail that barely widens past its base; its length is slightly longer than the head and the body. The body is almost uniformly grey, with a slightly lighter head. The tail is darker lavender. The chin, throat, and venter are yellowish white. The tail is unflattened below, and terminates in a very fine point. The male has 15-18 large femoropreanal pores in each side. **Behaviour:** Little known at present. **Habitat:** Largely unknown habitat requirements. Found in an old nypa palm house that was being torn down. **Range:** Recorded in Si Racha, Chonburi Province in southeastern Thailand. **Regional Range:** Endemic to Thailand.

61

62 Common Four-clawed Gecko *Gehyra mutilata*

Snout to Vent: 5.8 cm. Tail: 6.1 cm.

A 4-clawed gecko with a broad flattened tail behind the narrow vent. It has 15-20 lunate or femoropreanal pores on each side of the body; by comparison Fehlmann's Four-clawed Gecko only has 11 small subcircular femoropreanal pores on each side. Dorsal surface is brownish-lavender with two median rows of dim light spots. Chin, venter, and subcaudal region are whitish-yellow. Digits have very small webs between the fingers. The tail is flattened on the undersurface and denticulate on the sides. **Behaviour:** Nocturnal. Often seen climbing on walls and ceilings. **Habitat:** Commonly found in houses in villages, towns, and cities, as well as in forest. **Range:** Found throughout the country. **Regional Range:** India, Sri Lanka, Bangladesh, Myanmar, China, Vietnam, Laos, Cambodia, Thailand, Peninsular and Eastern Malaysia, Singapore, Brunei, Indonesia, Philippines, East Timor, Papua New Guinea, Australia, Oceania, and Mexico.

62

63 Fehlmann's Four-clawed Gecko *Gehyra fehlmanni*

Snout to Vent: 3.7 cm. Tail: 3.3 cm.

Like Common Four-clawed Gecko, it has a broad flattened tail and a narrower vent. Body coloration is light brown with black spots on the head, body limbs, and tail. The ventral surface is yellowish-white. It has 11 small subcircular femoropreanal pores on each side of the body; by comparison Common Four-clawed Gecko only has 15-20 lunate or femoropreanal pores on each side. The scales on the dorsal surface of head and body are not distinctly conical. The tail is denticulate on the sides. **Behaviour:** Nocturnal. Probably terrestrial, but little known. **Habitat:** Unknown. **Range:** Recorded in Kanchanaburi, Nakhon Ratchasima, and Nakhon Si Thammarat provinces. **Regional Range:** Endemic to Thailand.

63

Dwarf Geckos Genus Hemiphyllodactylus

Members of this genus are diminutive in size. They have four well-developed, clawed digits and one vestigial digit, which are not webbed. The digits bear widened lamellae. The dorsal scales are small and granular, and lack tubercles. Preanal and femoral pores are present. The pupil is vertical. Two species are found in Thailand.

KEY TO THE THAI *HEMIPHYLLODACTYLUS*

1 a Hind limb not reaching to more than halfway between the axilla and groin; no distinct postmentals; males with preanal and femoral pores — *H. typus*

b Hind limb reaching more than halfway between axilla and groin; postmentals distinct, males with preanofemoral pores in a single series — *H. yunnanensis*

Key modified from Taylor (1963).

64 **Malayan Dwarf Gecko** *Hemiphyllodactylus typus*

Snout to Vent: 4.6 cm. Tail: 4.0 cm.

Diminutive gecko marked with grey above and fine speckling of cinnamon brown. Head is yellowish-brown. The venter is yellowish-white. The hind limb does not reach more than halfway between the axilla and the groin, as in Yunnan Dwarf Gecko. Male has preanal and femoral pores; in Yunnan Dwarf Gecko these pores are in a single series. **Behaviour:** Nocturnal. Found on tree trunks. **Habitat:** Inhabits southeastern and peninsular monsoonal evergreen forest and mixed dipterocarp forest. **Range:** Two disjunct ranges, with records from Khlong Mao Nao, Trat Province; and records from Muang district, Ranong Province; from Khao Chong, Trang Province; and Hala Bala Wildlife Sanctuary, Yala Province. **Regional Range:** India, Sri

Lanka, Myanmar, Thailand, Peninsular and Eastern Malaysia, Singapore, Brunei, Indonesia, Philippines, and islands of Oceania.

64

65 Yunnan Dwarf Gecko ***Hemiphyllodactylus yunnanensis***

Snout to Vent: 4.1 cm. Tail: 3.7 cm.

Diminutive gecko with a median brown stripe and two lateral grey stripes. Arms and legs are mottled brown. The hind limb reaches more than halfway between the axilla and the groin. The male has preanofemoral pores in a single series. **Behaviour:** Nocturnal. **Habitat:** Found around human dwellings. **Range:** Northern Thailand, northeastern (western) Thailand, and from Ko Samui in Surat Thani Province. **Regional Range:** Myanmar, China, Vietnam, Laos, and Thailand.

65

Tokay Geckos Genus *Gekko*

Members of this genus have either free or partially webbed digits. The terminal phalanges of the outer four digits are slender and clawed. The inner digit is well-developed, but clawless. The dorsal scales are small and granular, and may be interspersed with larger tubercles. Five species are found in Thailand.

KEY TO THE THAI *GEKKO*

1 a	Nostril bordered by rostral; scales on snout distinctly larger than those on dorsum of body; 12-17 scansors underneath toe IV	**2**
a	Nostril separated from rostral by a scale; scales on snout as large as, or only slightly larger than, those on dorsum of body; 18-24 scansors underneath toe IV	**3**

2 a Adult male with pore series in preanal and femoral regions; dorsal tubercles distinctly enlarged and strongly convex, conical; 148-177 rows of scales around midbody *G. monarchus*

b Adult male with pores only in preanal region; dorsal tubercles not much enlarged or distinctly convex; 113-121 rows of scales around mid-body *G. petricolus*

3 a 16-19 rows of dorsal tubercles at mid-body *G. siamensis*

b 8-14 rows of dorsal tubercles at mid-body **4**

4 a Ear relatively large, its diameter 3.9-5.2% of snout to vent length; sub-caudals not widened *G. gecko*

b Ear relatively small, its diameter 2.1-3.8% of snout to vent length; sub-caudals distinctly widened *G. smithii*

Key modified from Ota and Nabhitabhata (1991).

66 Common Tokay Gecko *Gekko gecko*

Snout to Vent: 18.5 cm. Tail: 17.9 cm

One of the largest tokay geckos, reaching a length of more than 36 cm. Distinctive coloration, being bluish-grey to ultramarine above, with brick-red or orange-red spots and numerous grey-white spots. The large, bulky, head is covered with numerous rusty-red spots; it is also covered with small polygonal scales. The digits are entirely free with numerous undivided lamellae on the undersurface. It has 8-12 preanal pores on each side forming a continuous series. **Behaviour:** Nocturnal and territorial. During the daytime it rests in crevices. Food consists of large insects, including termites, moths, and beetles. **Habitat:** Found in all forest types. **Range:** Found throughout the country. **Regional Range:** India, Myanmar, China, Vietnam, Laos, Cambodia, Thailand, Peninsular and Eastern Malaysia, Singapore, Brunei, Indonesia, and the Philippines.

66

67 Spotted Tokay Gecko *Gekko monarchus*

Snout to Vent: 10.0 cm. Tail 10.4 cm.

A medium-sized tokay gecko, cream or fawn-coloured, with small, paired black spots running down the length of the back. It has 16-20 femoropreanal pores on each side of the body. The distal parts of the tail are banded. **Behaviour:** Nocturnal. Call is most frequently heard at twilight. **Habitat:** Inhabits lowland primary forests including peninsular monsoonal evergreen forest, Malayan mixed dipterocarp forest and peat swamp forest. **Range:** Found in the peninsular Thailand northward to Phang-nga Province. **Regional Range:** Thailand, Peninsular and Eastern Malaysia, Singapore, Brunei, Indonesia, and the Philippines.

67

68 Sandstone Tokay Gecko *Gekko petricolus*

Snout to Vent: 9.8 cm. Tail: 11.4 cm.

Separated from other tokay geckos by head and body coloration. The head is lavender-grey, and the body is greenish-yellow with two longitudinal rows of small circular yellowish-white spots alternating with small blackish spots which may be absent. The underside of body and tail is yellowish or dirty-white. The tail is slender and depressed. It has 9 or 10 preanal pores; but no femoral pores. **Behaviour:** Inhabits cracks and crevices in sandstone rocks. **Habitat:** Found in horizontal rocks in sandstone hills. **Range:** Found in northeast Thailand. **Regional Range:** Laos and Thailand.

68

69 Siamese Tokay Gecko *Gekko siamensis*

Snout to Vent: 13.4 cm. Tail: 11.4 cm.

Large-headed species covered with small polygonal scales. Grey to ultramarine above, with brick-red or orange-red spots and numerous grey-white spots. It has green eyes. **Behaviour:** Nocturnal. Found among fallen sandstone rocks and boulders. **Habitat:** Found in mixed deciduous forest. **Range:** Recorded from Petchabun Province and Pak Chong, Nakhon Ratchasima Province. **Regional Range:** Endemic to Thailand.

69

70 Smith's Tokay Gecko *Gekko smithi*

Snout to Vent: 17.2 cm. Tail: 15.5 cm

One of the largest geckos. It lacks the reddish spots found in the Common Tokay Gekko. It has green eyes. Dorsal surface is grey with eight transverse rows of white tubercules. Ventral surface is greyish-white. Tail is banded with light and dark grey. **Behaviour:** Nocturnal. Arboreal, often found high in tall trees, foraging on tree trunks. In daytime, found in tree holes. **Habitat:** Found in peninsular monsoonal evergreen forest and Malayan mixed dipterocarp forest. **Range:** Found in peninsular Thailand. **Regional Range:** Myanmar, Thailand, Peninsular and Eastern Malaysia, Singapore, Brunei, and Indonesia.

70

Parachute Geckos Genus *Ptychozoon*

This enigmatic genus has strongly webbed digits, a skin-flap on the side of the head, another flap on the sides of the body and the tail is expanded into numerous separate scallops, sometimes with an enlarged terminal flap.

They have dorsal scales intermixed with larger tubercles. They inhabit climax monsoonal forest. These species are nocturnal, hiding by day in dark crevices or inside trunks. The pupil is vertical. Males have preanal and femoral pores. Four species are found in Thailand.

KEY TO THE THAI *PYTCHOZOON*

1 a Three transverse wavy bands in the axilla-groin region; flat dorsal tubercles present — *P. trinotaterra*

b Four transverse bands in the axill-groin region; dorsal tubercles if present are convex or speculate — **2**

2 a Male with preanal and femoral pores separated, preanal pores 10-11; no enlarged dorsal tubercles; tail tapering gradually to tip, scallops on tail smaller and directed slightly backward; no enlarged tubercles — *P. horsfieldii*

b Male with continuous preanal and femoral pores — **3**

3 a Dorsal body scales intermixed with larger tubercles; preanal pores number 14-32; tail widened at tip into a flap, the scallops wider than in *P. horsfieldii* and placed at right angles to axis of tail — *P. kuhli*

b Males with from 15 to 26 preanal pores; no enlarged tubercles intermixed with dorsal body scales; scallops narrow directed slightly backward; tail not forming a widened flap at tip — *P. lionotum*

Key modified from Taylor (1963); Brown (1999).

71 Common Parachute Gecko *Ptychozoon lionotum*

Snout to Vent: 8.0 cm. Tail: 8.1 cm

Tail does not form a widened flap at the tip. No enlarged tubercles among dorsal body scales, unlike Horsfield's and Kuhl's Parachute Gecko. Males have 16-25 preanal pores. Dull grey-lavender with four narrow transverse brown marks. Chin, throat, and breast are whitish. Tail with several spots or bands. **Behaviour:** Nocturnal and arboreal, often found at high in tall trees. **Habitat:** Found in a wide range of forest habitats. **Range:** Found in northern, northeastern (west), western, and peninsuar Thailand. **Regional Range:** Myanmar and Thailand.

71

72 Three-banded Parachute Gecko *Ptychozoon trinotaterra*

Snout to Vent: 7.1 cm. Tail: 15.5 cm

Smaller-sized than Common and Kuhl's Parachute Gecko. Separated from all Parachute Lizards by presence of four dark transverse bands in the axilla-groin region; all other species have four bands. Differs from Horsfield's Parachute Gecko by presence of an enlarged tail terminus. Further differs from Common Parachute Gecko by limited terminal flap expansion. **Behaviour:** Nocturnal. Arboreal, often found up high in tall trees. **Habitat:** Found in dry evergreen forest. **Range:** Found in northeastern Thailand. Recorded from Sakaerat, Nakhon Ratchasima Province; and Ubon Ratchathani Province. **Regional Range:** Vietnam and Thailand.

72

73 Kuhl's Parachute Gecko *Ptychozoon kuhli*

Snout to Vent: 9.4 cm. Tail: 6.3 cm

Dorsal body scales interspersed with larger flat, rounded tubercles. Tail also widens at tip into a flap. The scallops are wider than in Horsfields' Parachute Lizard and are at right angles to the axis of the tail. Head nearly uniform grey-brown above. Body brownish to lavender-grey with a series of blackish spots across nape of neck. The chin, throat, and venter are dirty white. It has 20-25 preanal pores. **Behaviour:** Nocturnal. Found

73

high up in the canopy. **Habitat:** Inhabits peninsular monsoonal evergreen forest and Malayan mixed dipterocarp forest, including disturbed forest. **Range:** Found in lower peninsular Thailand. Records from Nakhon Si Thammarat, Pattani, Narathiwat, and Satun provinces. **Regional Range:** India (Andamans), Thailand, Peninsular and Eastern Malaysia, Singapore, Brunei, and Indonesia.

74 Horsfield's Parachute Gecko *Ptychozoon horsfieldii*

Snout to Vent: 7.4 cm. Tail: 15.5 cm

Tail characteristically tapers gradually to a tip. Scallops on tail smaller and directed somewhat backward. Males have separated preanal and femoral pores; the preanals number 10-11 and the femoral pores 8-11. **Behaviour:** Nocturnal and arboreal. Found high up in the canopy, rarely descending to the lower strata. **Habitat:** Found in Malayan mixed dipterocarp forest. **Range:** Recorded from Yala Province. **Regional Range:** Thailand, Peninsular Malaysia, and Indonesia (Kalimantan and Sumatra).

74

Agamid Lizards Family Agamidae

The agamids are largely confined to Asia, Australia, and Africa. The teeth are acrodont, varying in shape and not replaced. The tongue is thick. Femoral pores are absent in most species, except the genera *Leiolepis, Physignathus,* and *Mantheyus*. Eleven genera, comprising 28 species, are represented in Thailand, namely the gliding lizards *Draco;* the tree lizards *Acanthosaura;* the anglehead lizards *Gonocephalus;* the crested lizards *Bronchocela, Pseudocalotes,* and *Calotes;* the water dragon *Physignathus;* the Phu Wua lizard *Mantheyus phuwuanensis,* the earless lizard *Aphaniotis*; the butterfly lizards *Leiolepis;* and the long-tailed lizard *Takydromus.*

KEY TO THAI GENERA OF THE AGAMIDAE

1 a	Body depressed	**2**
b	Body compressed	**4**

2	**a**	Five or six elongated ribs supporting lateral dermal expansions serving as wings for gliding flight; a longer or shorter gular projection (dewlap) at least in males; arboreal	*Draco*
	b	Ribs not supporting a dermal expansion for gliding flight; sometimes a small gular pouch but no typical dewlap	**3**
3	**a**	Gular with two transverse skin folds; tail depressed, thickened at base	*Leiolepis*
	b	Gular with U-shaped skin fold; tail compressed	*Mantheyus*
4	**a**	Femoral pores (at least in males) 4-8 on each side; tail strongly compressed, covered with small keeled scales above, and much larger, more strongly keeled one below; hindleg reaches to eye or tip of snout	*Physignathus*
	b	Femoral pores not present	**5**
5	**a**	Tympanum hidden; no gular fold or pouch; inside of mouth blue	*Aphaniotis*
	b	Tympanum exposed; mouth lining not blue; small gular folds sometimes present	**6**
6	**a**	A strong spine arising behind orbit, or on side of neck	*Acanthosaura*
	b	No strong spine arising behind orbit, or on side of neck	**7**
7	**a**	Dorsal scales unequal, intermixed with few or many larger irregular scales; a high dorsal crest in males, the crest slightly lower in females	*Gonocephalus*
	b	Dorsal scales nearly uniform, not intermixed with larger scales; body scales regularly arranged	**8**
8	**a**	Lateral scales in rows pointing backwards and downward; head without enlarged compressed scales behind the orbit	**9**
	b	Lateral scales in rows pointing backward and upward; head with enlarged compressed scales behind the orbit	*Calotes*
9	**a**	A cheek skin fold is absent	*Pseudocalotes*
	b	A cheek skin fold is present	*Bronchocela*

Key modified from Taylor (1965).

Gliding Lizards Genus *Draco*

Members of this genus can glide from one tree to another. To facilitate gliding, they have 4-7 elongate ribs passing through the body wall, supporting a wide membrane (patagium) continuous with body skin that may be folded parallel to the body or expanded like a fan by rib movements. Nine species are found in Thailand.

KEY TO THE THAI *DRACO*

1	**a**	Nostril directed outward or obliquely upward; a longitudinally enlarged scale, sometimes compressed to a triangular thorn, may be present on the supraciliary edge, just above and posterior to the centre of the eye	**2**
	b	Nostril directed upward; no longitudinally enlarged scale on the supraciliary edge	**4**

2 a	Number of ribs supporting the patagium 6 or 7	*D. sumatranus*
b	Number of ribs supporting the patagium 5	**3**
3 a	Number of dorsals usually more than 160; number of supralabials 8-13; males with a caudal crest	*D. fimbriatus*
b	Number of dorsals usually less than 160; number of supralabials 8-11; males without a caudal crest.	*D. maculatus*
4 a	Number of ribs supporting the patagium 6 or 7	*D. quinquefasciatus*
b	Number of ribs supporting the patagium 5	**5**
5 a	Gular pouch in males and females without enlarged scales; males without a nuchal fold; patagium above usually dark, with light spots	**6**
b	Gular pouch at least in males with enlarged scales; males with a nuchal fold; patagium above usually brown or grey with or without dark markings	**7**
6 a	Gular pouch in males coal black; in females grey; number of dorsals 164-218; patagium dark brown with distinct, light spots	*D. melanopogon*
b	A coal black spot at the base of the gular pouch in males; in females a grey brown spot at the base of the gular pouch; number of dorsals 148-184; patagium usually brown with large, light spots; sometimes forming bands	*D. haematopogon*
7 a	Males without a caudal crest; females without enlarged scales on the gular pouch; snout-vent length less than 7.8 cm	*D.taeniopterus*
b	Males with a more or less distinct caudal crest; females with enlarged scales on the tip of the gular pouch; snout-vent length in adults more than 7.5 cm	**8**
8 a	Patagium in males brown, sometimes with indistinct bands; in females with distinct bands; have a black dividing line on the border between the ventral parts of the lateral pouch and the gular pouch; gular pouch not widened distally in males	*D. blandfordii*
b	Patagium in both sexes brown with dark spots or irregular band-like markings; no black dividing line between the ventral parts of the lateral pouch and the gular pouch; gular pouch distally widened to a rounded flap	*D. obscurus*

Key modified from Muster (1983).

75 Sumatran Gliding Lizard *Draco sumatranus*

Snout to Vent: 8.5 cm. Tail: 12.6 cm.

One of the smaller *Draco* species. Underside of wing membrane is a diagnostic greyish-ultramarine blue with numerous black spots. Upper side of wing membrane is brown with the outer posterior parts dark brown with lighter spots; the bands are not clearly recognizable. Wing membrane has six ribs. It has no caudal crest. Body is variegated bronze-brown, olive, black, and fawn above with some metallic iridescence. Triangular gular pouch (dewlap) is ultramarine blue at base with the distal parts yellowish or fleshy pink; it is pointed but rarely longer than the head. Nostrils point laterally.

No thorn-like scale on supraciliary edge. The male has a nuchal crest. Tail is weakly banded. **Behaviour:** Arboreal, moving from one tree to another by spreading gliding membrane. Usually stays in the upper and middle storeys of the canopy, rarely descending to lower levels. **Habitat:** Found in lowlands in peninsular monsoonal evergreen forest and Malayan mixed dipterocarp forest, as well as plantations and gardens. Prefers dense areas of forest growing adjacent to large streams. Generally scarce. **Range:** Peninsular Thailand, extending up the Tenasserim range. Recorded in Kanchanaburi, Nakhon Si Thammarat, Trang, Narathiwat, and Yala provinces. **Regional Range:** Thailand, Peninsular and Eastern Malaysia, Singapore, Brunei, Indonesia, and the Philippines.

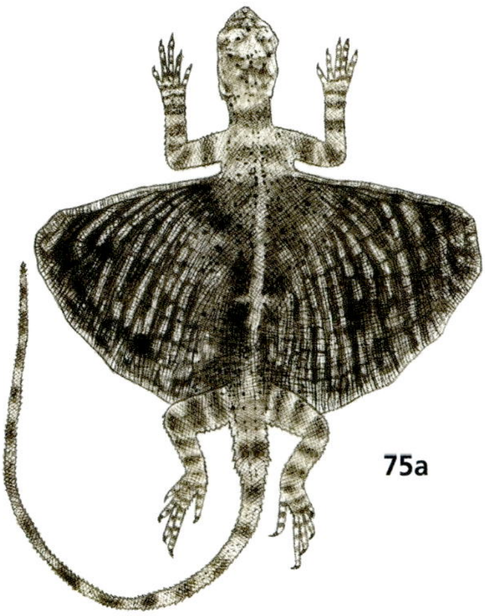

75a

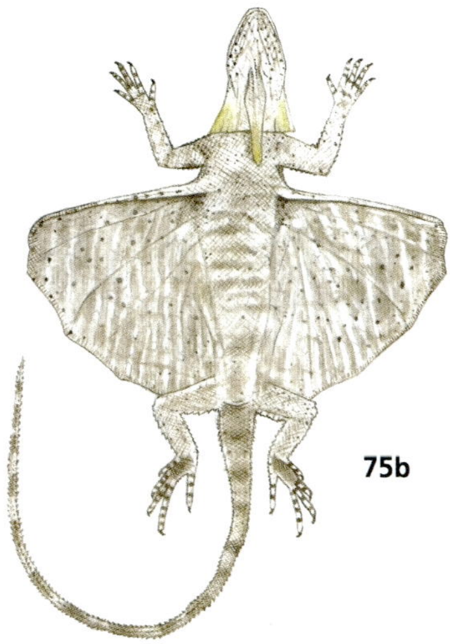

75b

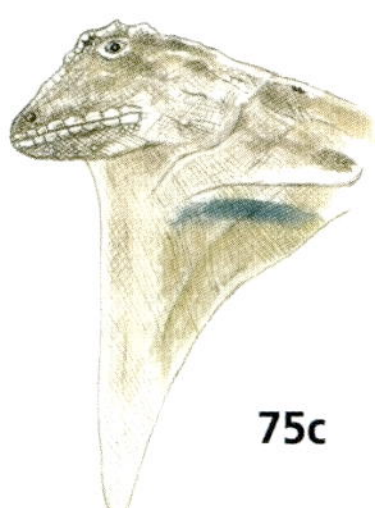

75c

76 Orange-winged Gliding Lizard *Draco maculatus*

Snout to Vent: 8.6 cm. Tail: 11.4 cm.

A small *Draco* species, with three subspecies occurring in Thailand. The dorsum is pale-blue or olive-grey. Wing membranes of male usually yellowish, orange, or orange-red with numerous black spots on the ventral surface. It has five ribs. The gular pouch is elongate, tapering near base, broad at the tip and covered with small scales. In *D. m. maculatus* (all regions apart from north-east) and *D. m. divergens* (from Chiang Mai only) a blue spot is found on each side of the base of the gular pouch. In *D. m. haasei* (from continental

Thailand) the base of the gular pouch is yellowish-white. Nostrils point laterally. A low thorn-like scale on the supraciliary edge. **Behaviour:** Diurnal and arboreal, actively foraging mainly at dawn and dusk by moving up and down tree trunks. When the sunlight is strong, an individual will bask on a tree trunk for up to half an hour. The female will bury her eggs at the base of a tree which will be warmed by the sunlight. **Habitat:** Occurs in open forest habitats, including dry dipterocarp forest, forest edges, and rubber plantation. Avoids dense vegetation. Frequently found in rubber plantations and orchards. **Range:** Found in forest habitats throughout Thailand. **Regional Range:** Myanmar, China, Vietnam, Laos, Cambodia, Thailand, and Peninsular Malaysia.

76a

76b

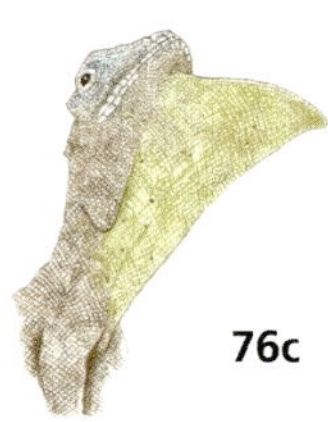

76c

77 Orange-bearded Gliding Lizard *Draco fimbriatus*

Snout to Vent: Up to 11.6 cm. Tail: Up to 19.1 cm.

A large *Draco*. The dorsum and patagium are greyish-brown, with grey markings. Male has a distinct nuchal crest and a weakly defined caudal crest; absent in female. Gular pouch (dewlap) is elongate, gradually tapering to a broad tip. Male wattle and dewlap ranges from saffron to orange with indistinct radiating grey lines. Head is brownish-olive above with black punctuations. Has an inverted Y-shaped series of scales usually evident on snout. Nostrils point laterally or slightly obliquely. A low thorn-like scale on supraciliary edge. Underside of wing is greyish with fairly large spots on distal portion. Five ribs in wing membrane. Laterally compressed tail is banded brownish and grey at base, becoming clay and blackish more distally. **Behaviour:** Arboreal, moving from one tree to another by spreading gliding membrane. Diurnal. Shows strong territorial behaviour. **Habitat:** Found in peninsular monsoonal evergreen forest, and Malayan mixed dipterocarp forest. **Range:** Southern provinces in peninsular Thailand. Recorded in Nakhon Si Thammarat, Trang, Pattani, Narathiwat, and Yala provinces. **Regional Range:** Thailand, Peninsular and Eastern Malaysia, Singapore, Brunei, and Indonesia (Sumatra).

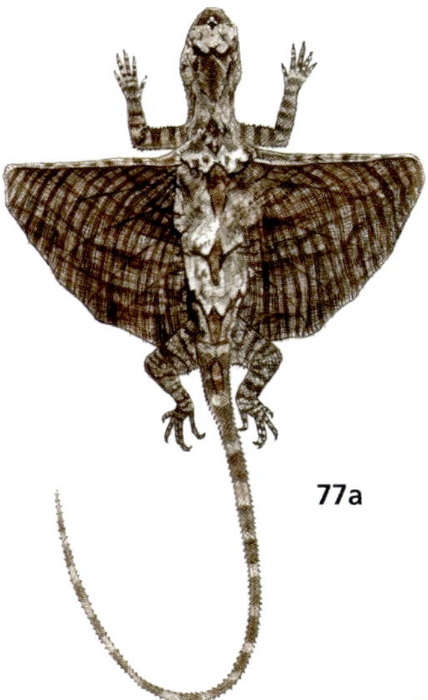

77a

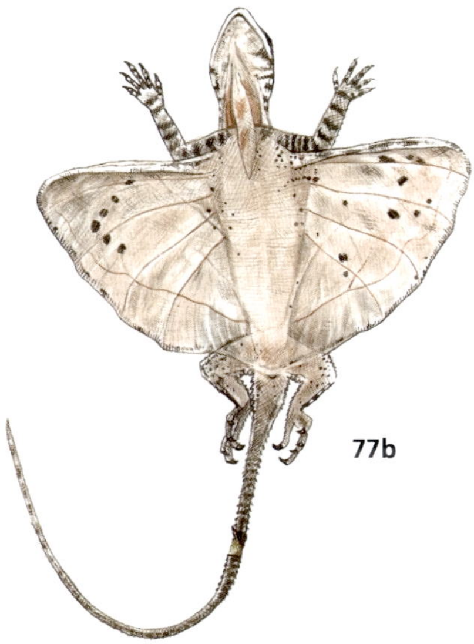

77b

77c

78 Five-banded Gliding Lizard *Draco quinquefasciatus*

Snout to Vent: 10.5 cm. Tail: 16.6 cm.

A large *Draco*. Upper surfaces of wings are characteristically red with five regular curved black bands of varying width; each band has a transverse row of small white spots. Undersides of wings have narrow transverse blackish bands. A black band crosses underneath the neck. Dewlap is distally greyish-brown. Head, neck, and upper body are reddish-brown dotted with black. Nostrils point dorsally. No thorn-like scale on supraciliary edge. No caudal crest. The tail is banded grey and brown. **Behaviour:** Arboreal and diurnal. **Habitat:** Found in peninsular monsoonal evergreen forest, peat swamp forest, and Malayan mixed dipterocarp forest. **Range:** Peninsular Thailand. Recorded in Surat Thani, Trang, Pattani, Narathiwat, and Yala provinces. **Regional Range:** Thailand, Peninsular and Eastern Malaysia, Singapore, Brunei, and Indonesia (Sumatra).

78a

78b

78c

79 Red-bearded Gliding Lizard *Draco haematopogon*

Snout to Vent: 9.8 cm. Tail: 11.4 cm.

A moderately sized species. The body is usually greenish or greyish above; the underside is greenish. The gliding membrane is blackish above with yellowish spotting. The lower surface of the membrane is green or grey. The absence of maroon or magenta in the outer area of the gliding membrane separates the species from similar Barred Gliding Lizard. The membrane has five ribs. Males have a large coal-black spot at the base of the gular pouch (dewlap), in females this spot is grey-brown. Dewlap is about one-and-a-half times as long as head, triangular, tapering gradually, and covered with small scales. Nostrils point dorsally. No thorn-like scale on supraciliary edge. No nuchal crest. **Behaviour:** Arboreal, gliding between trees. **Habitat:** Peninsular monsoonal evergreen forest. **Range:** Only recorded on Koh Phangan, Surat Thani Province in Peninsular Thailand. **Regional Range:** Thailand, Peninsular and Eastern Malaysia, and Indonesia (Sumatra).

80 Black-bearded Gliding Lizard *Draco melanopogon*

Snout to Vent: 8.5 cm. Tail: 11.4 cm.

Small to moderately sized gliding lizard. Head, back, limbs, and base of tail are variegated green and olive. Venter is dull yellow with reddish-brown flecks. The wing membranes are characteristically coloured black with small distinct yellow or yellowish-brown speckles. The tail is banded black and reddish-brown. The male has a diagnostic coal-black dewlap, which is easily observed in the field; in females it is grey. Toward the base of the dewlap the posterior area is grey, ultramarine, or pure white; this dewlap is much longer than the head. Nostrils are directed upward. No thorn-like scale on supraciliary edge. **Behaviour:** Arboreal and diurnal. Tends to prefer lower forest strata. Prefers dim light conditions. **Habitat:** Found in riverine and dense streamside vegetation within peninsular monsoonal evergreen forest and Malayan mixed dipterocarp forest. **Range:** Peninsular Thailand, south of the Isthmus of Kra. Recorded in Surat Thani, Krabi,

80a

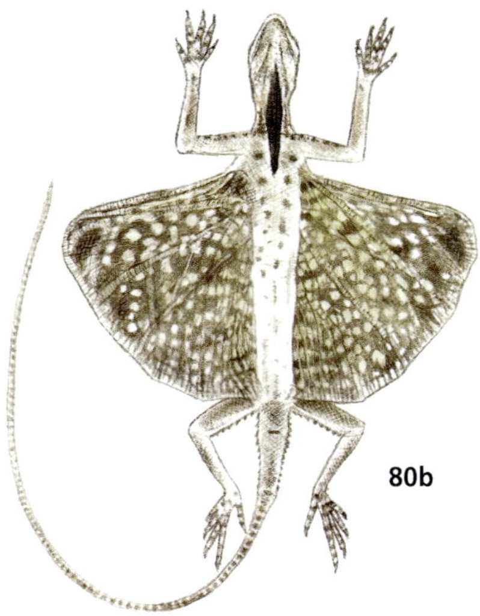

80b

80c

Nakhon Si Thammarat, Phattalung, Trang, Narathiwat, and Yala provinces. **Regional Range:** Thailand, Peninsular and Eastern Malaysia, Singapore, and Indonesia (Sumatra).

81 Barred Gliding Lizard *Draco taeniopterus*

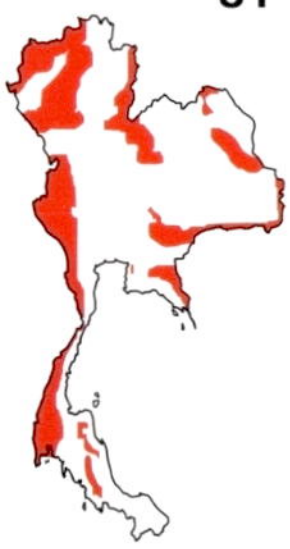

Snout to Vent: 7.8 cm. Tail: 15.2 cm.

One of the smallest gliding lizards. Head is dark olive above. Body is greyish-olive or bronze-coloured. Underside is immaculate with a yellow wash. Wings are olive above with 4-5 irregular, jet-black bands. The outer posterior part of the wing is reddish or magenta; this coloration is absent in the Red-bearded Gliding Lizard. Underside of wing membrane is yellow. Distal part of dewlap pale-green or olive, rounded and covered with large translucent scales. Basal part of dewlap and wattle is orange-red or carmine red. No thorn-like scale on supraciliary edge. **Behaviour:** Arboreal. Diurnal. **Habitat:** Prefers forest edge habitats in a wide range of forest types. **Range:** Found throughout Thailand, apart from extreme southern peninsular Thailand. Recorded in Chiang Mai, Uthai Thani, Kanchanaburi, Nakhon Ratchasima, Chanthaburi, Trat, Chumphon, Ranong, Surat Thani, Phuket, Krabi, and Trang provinces. **Regional Range:** Myanmar, Cambodia, and Thailand.

81a

81b

81c

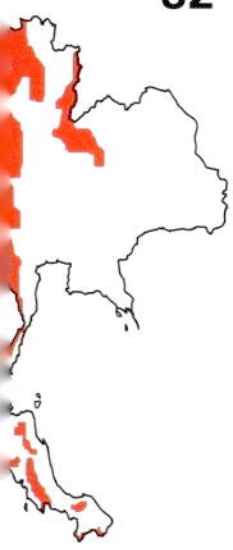

82 Blandford's Gliding Lizard *Draco blandfordii*

Snout to Vent: 13.4 cm. Tail: 25.6 cm.

The largest gliding lizard. Above, body is variegated greyish or greyish-olive. Hind limbs are usually shorter than the distance between the limbs. Tail is banded brown and grey. Wings are olive, olive-grey, or olive-yellow above; in male with fine, lighter, parallel, longitudinal or slightly curving lines; in female with more or less well-defined cross bars which may be restricted to a black reticulation at the outer margin. A few maroon or reddish flecks are present on the outer posterior part of wing. Underwings are uniformly yellow. Venter is whitish. Wattle ranges from orange to scarlet-red. Distal

82a

82b

82c

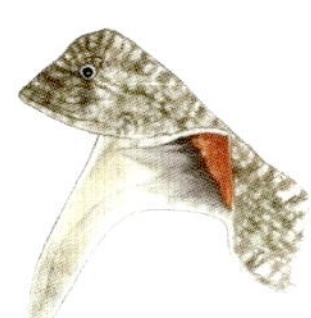

82d

two-thirds of dewlap covered with large translucent scales. Nostrils point obliquely. No thorn-like scale on supraciliary edge (above eye on top of head). **Behaviour:** Arboreal and diurnal. **Habitat:** Prefers hilly habitats, although young are found at lower elevations. Occurs in mixed deciduous and dry evergreen forests, peninsular monsoonal evergreen forest, and Malayan mixed dipterocarp forest. **Range:** Northern, western, and peninsular Thailand. Recorded in Chiang Rai, Chiang Mai, Phrae, Loei, Uthai Thani, Kanchanaburi, Prachuap Khiri Khan, Chumphon, Ranong, Phuket, Phang-nga, Krabi, Nakhon Si Thammarat, Trang, Songkhla, Pattani, Narathiwat, Satun, and Yala provinces. **Regional Range:** Myanmar, China, and Thailand.

83 Dusky Gliding Lizard *Draco obscurus*

Snout to Vent: 8.4 cm. Tail: 14.3 cm.

Medium-sized *Draco*. Dorsum is greyish-brown with scattered brown flecks. Has a relatively large head. Wings are black with fine, broken grey streaks. Six ribs in wings. Tail is banded with light and dark grey. Dewlap of male is chrome yellow. Has a very distinct nuchal crest. Caudal crest has rather

83a

83b

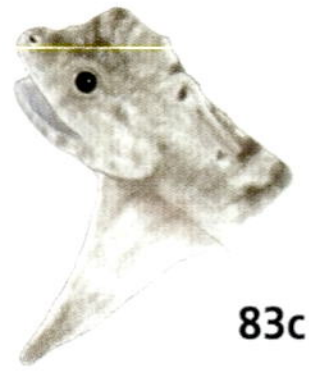

83c

long vertical pointed scales. Nostrils point dorsally. No thorn-like scale on supraciliary edge. **Behaviour:** Arboreal, often seen gliding accross streams. Spends the night only 1-1.5 m above the ground. **Habitat:** Prefer stream habitats, in peninsular monsoonal evergreen forest, and Malayan mixed dipterocarp forest. **Range:** Peninsular Thailand. Found in Chumphon, Nakhon Si Thammarat, Trang, Pattani, Narathiwat, Satun, and Yala provinces. **Regional Range:** Thailand, Peninsular and Eastern Malaysia, and Indonesia (Sumatra).

Tree Lizards Genus *Acanthosaura*

This group is characterized by a somewhat compressed body, usually with a sharp dorsal ridge and a dorsal crest. The dorsal scales are unequal. A gular pouch may be present. A fold across the shoulders is usually conspicuous. Three species occur in Thailand.

KEY TO THE THAI *ACANTHOSAURA*

1 a No diamond-shaped black mark on neck and shoulders; no Y-shaped series of scales on snouts; tympanum large, naked; dorsal and nuchal crests separated by a small distema 2-3 mm in length; tail base swollen in males; a small nuchal pouch — *A. armata*

b A diamond-shaped black mark on the dorsum of neck — **2**

2 a Spines of nuchal and dorsal crests separated by a distema of 7-8 mm, the intervening space covered with small equal scales similar to upper body scales; nuchal black spot with arms reaching down on sides connecting with diagonal black spot — *A. crucigera*

b Spines of the nuchal and dorsal crests very close together but there is actually a separation since the spines at end of nuchal and beginning of nuchal may be shorter; postorbital and nuchal spines not half as long as orbit; no gular pouch — *A. lepidogaster*

Key modified from Taylor (1963).

84 Horned Tree Lizard *Acanthosaura armata*

Snout to Vent: 11.8 cm. Tail: 16.6 cm.

Large tree lizard with a uniformly brownish-fawn body with a darker brown, triangular-shaped head. Distinguished from other tree lizards by having obvious long spines behind the orbit and on the side of the neck, and by the absence of a diamond-shaped black mark on the neck and shoulders. Tympanum is also characteristically large and naked. No Y-shaped series of scales on snout. The limbs are greyish-brown and lightly banded. The tail is greyish-brown; in males the base is strongly swollen. **Behaviour:** Mainly terrestrial, spending most of the time on the ground feeding on slugs and earthworms. May climb trees up to about 2 m from the ground, and is often found resting in low branches. It rests overnight on low branches concealed amongst leaves. **Habitat:** Found in peninsular monsoonal evergreen forest and Malayan mixed dipterocarp forest. **Range:** Peninsular Thailand. Recorded in Nakhon Si Thammarat, Pattani, Narathiwat, and Yala provinces. **Regional Range:** Thailand, Peninsular Malaysia, Singapore, and Indonesia (Sumatra).

84

85 Scale-bellied Tree Lizard *Acanthosaura lepidogaster*

Snout to Vent: 9.1 cm. Tail: 11.8 cm.

A large tree lizard with a dark blackish-brown head and a greyish-brown body and limbs. It has both a nuchal and a dorsal crest, whose spines are very close together but there is actually a separation since the spines at the end of nuchal crest and beginning of dorsal crest may be shorter; in the Cross-bearing Tree Lizard there is a small gap (diastema) of 7-8 mm. The outer part of tympanum is covered in small scales. The short spines above the eyes—being less than half the diameter of the orbit—and the presence of a diamond black mark on the dorsum of the neck separates the species from Horned Tree Lizard. Also usually has a Y-shaped series of enlarged scales on snout. There is no gular pouch. **Behaviour:** Mainly terrestrial and solitary, though may form pairs during the breeding season. Feeds in the leaf litter for termites and insect larvae, with chicken-like pecking behaviour. Normally territorial. **Habitat:** Found in sub-montane, hill evergreen, and mixed deciduous forest. **Range:** Northern and northeastern (west) Thailand. Recorded in Chiang Mai, Lampang, Tak, Loei,

85

Petchabun, and Chaiyaphum provinces. **Regional Range:** Myanmar, China, Vietnam, Laos, Cambodia, and Thailand.

86 Cross-bearing Tree Lizard *Acanthosaura crucigera*

Snout to Vent: 10.3 cm. Tail: 14.6 cm.

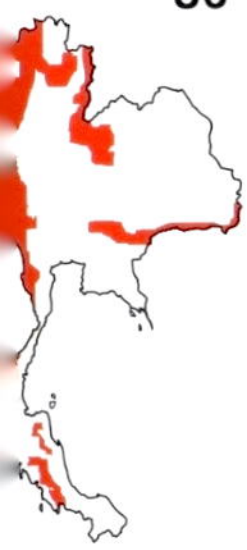

A large lizard with a deep-black reticulum enclosing yellowish or yellowish-brown spots. The tail is banded with dark and light brown. The spines of nuchal crest and dorsal crest separated by a diastema of 7-8 mm; in the similar Scale-bellied Tree Lizard this gap is only 2 mm wide. The spines behind the head and on the side of neck often as long as diameter of eye. Distinguished from Horned Tree Lizard by the presence of diamond-shaped black mark on dorsum of neck and a naked tympanum. **Behaviour:** Diurnal. Terrestrial and territorial, living on the forest floor where it searches for earthworms and insects. When threatened, climbs adjacent trees trunks. **Habitat:** Found in sub-montane and montane forest and dry evergreen forest at low altitudes. **Range:** Found in forested areas throughout Thailand. **Regional Range:** Myanmar, Laos, and Thailand.

86

Anglehead Lizards Genus *Gonocephalus*

This genus is characterized by the presence of a dorsal crest. They have strongly compressed bodies and compressed tails. A gular pouch is usually evident in the males. They are most usually arboreal, often being found high in the canopy. Three species occur in Thailand, confined to the south of the country.

KEY TO THE THAI *GONOCEPHALUS*

1	**a**	Superciliary border strongly elevated; dorsal crest almost as high as nuchal crest; a row of five enlarged scales on the sides	*G. abbotti*
	b	Superciliary border not strongly elevated	**2**
2	**a**	Enlarged scales scattered over sides; nuchal and dorsal crests continuous, very high in the male; dorsal, gular, and ventral scales strongly keeled	*G. bellii*
	b	No enlarged scales on sides; nuchal crest very high in the male, composed of long spines united together and free only at tips; a notch between nuchal and dorsal crests; gular and ventral scales smooth	*G. grandis*

Key modified from Taylor (1963).

87 Abbott's Anglehead Lizard *Gonocephalus abbotti*

Snout to Vent: 14.0 cm. Tail: 30.0 cm.

A large species with characteristic strongly elevated ridges above the eyes. The nuchal crest and dorsal crest are continuous, extending down the tail, with all the scales of even size. Mature males are brick red; females are bright green with black lines radiating from the eye. Both the body and tail are strongly depressed. The tail has eight dark bands separated by pinkish fawn bands. **Behaviour:** Arboreal, found high up in the canopy. Reportedly may stay in one place for several days. **Habitat:** Found in peninsular monsoonal evergreen forest and Malayan mixed dipterocarp forest. **Range:** Peninsular Thailand. Recorded in Trang, Narathiwat, and Yala provinces. **Regional Range:** Thailand and Peninsular Malaysia. **International Conservation Status:** IUCN: Data Deficient.

87

88 Bell's Anglehead Lizard *Gonocephalus bellii*

Snout to Vent: 13.8 cm. Tail: 32.8 cm.

A large lizard with a continuous nuchal and dorsal crest. The ridges over the eyes are not strongly elevated. Gular sac is well-developed. Dorsal, gular and ventral scales are strongly keeled. The top and sides of the head are olive. The throat pouch is bluish, the scales with black bases. The sides are variegated olive and lavender with some blackish flecking. The enlarged scales are often blue. Both the dorsal and nuchal crests are bluish to ultramarine with black at the base of the crest above shoulders. The tail is banded grey-olive and whitish. A blackish area in front of shoulders. The venter and under the thighs are amber. **Behaviour:** Arboreal, being found in high trees. **Habitat:** Found in peninsular monsoonal evergreen forest. **Range:** Peninsular Thailand. Two records, from the Surat Thani - Nakhon Si Thammarat border and Narathiwat Province. **Regional Range:** Thailand, and Peninsular and Eastern Malaysia.

88

89 Great Anglehead Lizard *Gonocephalus grandis*

Snout to Vent: 15.5 cm. Tail: 39.6 cm.

The largest Anglehead Lizard, with a compressed body and tail. In males the ribbon-like nuchal crest may be very high, composed of long spines united together and free only at the tips. The dorsal crest, which is half as tall as the nuchal crest, extends along the back to a position level with the vent. There are no enlarged scales on the flanks, as are present on Bell's Anglehead Lizard. The eyebrow ridge is not strongly elevated. Generally lavender-brown when at rest with bluish-grey bands on the limbs and digits. When stressed, males turn olive-green; females and juveniles turn reddish-brown. Chin and throat cream-coloured. Tail has about 20-22 alternating bluish-white and brown bands. **Behaviour:** Arboreal, found high up in the canopy. Young may be found close to streams among overhanging vegetation and even descend to the ground to feed around tree bases. Up to six eggs laid in a clutch. **Habitat:** Found in Malayan mixed dipterocarp forest. **Range:** Peninsular Thailand. Recorded in Songkhla, Pattani, Narathiwat, Satun, and Yala provinces. **Regional Range:** Thailand, Peninsular and Eastern Malaysia, and Indonesia (Sumatra).

89

Crested Lizards Genus *Bronchocela*

Body compressed for arboreal habits. Scales on sides of the body point backward and downward; in *Calotes* and *Pseudocalotes* these scales point backward and upward. The dorsal scales are regular and uniform. A dorsal crest is usually present, higher in males. They are long-headed and slender-limbed. Tail is often 2-3 times length of head and body; in the adult male it is not markedly swollen at the base. Femoral and preanal pores absent. In Thailand 2 species are present.

KEY TO THE THAI *BRONCHOCELA*

1 a	Has 6-10 upper dorsal scales pointing backward and upward; fourth finger as long as fifth toe	*B. cristellata*
b	Has only 2-4 upper dorsal scales pointing backward and upward; fourth finger much longer than fifth toe.	*B. smaragdina*

Key modified from Taylor (1963).

90 Green Crested Lizard *Bronchocela cristatella*

Snout to Vent: 12.2 cm. Tail: 45.4 cm.

Normally a brilliant green-coloured lizard, though the colour may change. The nuchal crest composed of about 10 small, erect spines, and is continuous with a serrated ridge which extends onto the dorsum, but does not reach the tail. It has slender, elongate arms and legs. The tympanum is quite large, being greater than the half the orbit diameter. The male has a small gular pouch, which is absent in females. Six to 10 upper dorsal scale-rows point backward and upward. The small dorsal scales are keeled. The ventrals are strongly keeled, being 3-5 times larger than the dorsals. **Behaviour:** Arboreal, preferring streamside trees and shrubs. At night and during heavy rain may change colour from brilliant green to almost black. Young stay in low, thick vegetation near streams. **Habitat:** Found in semi-evergreen forest, peninsular monsoonal evergreen forest, and Malayan mixed dipterocarp forest. **Range:** Found in western and peninsular Thailand. Recorded in Kanchanaburi, Chumphon, Surat Thani, Nakhon Si Thammarat, Trang, Pattani, and Yala provinces. **Regional Range:** India (Nicobars), Myanmar, Thailand, Peninsular and Eastern Malaysia, Singapore, Philippines, Indonesia (Sumatra, Java and Sulawesi), and Papua New Guinea.

90

91 Smaragdine Crested Lizard *Bronchocela smaragdina*

Snout to Vent: 9.5 cm. Tail: 30.0 cm.

Emerald green above and greenish-white below. Differs from Green Crested Lizard in having no dorsal crest, while nuchal crest is composed of small erect scales only. A white or yellowish stripe runs from the armpit along the lower flanks of the body and terminates at the base of the tail. The upper 2-4 dorsal scale-rows only point backward and upward or straight backward. Gular scales are larger than in Green Crested Lizard. It has 50-53 scale-rows around the middle of the body. **Behaviour:** Arboreal and usually solitary. Found high up in shrubs and trees, particularly on the tips of outstretched branches, overhanging waterways. **Habitat:** Found in southeastern monsoonal evergreen forest. **Range:** Only recorded in Khao Sao Doi Wildlife Sanctuary in Chanthaburi Province, southeastern Thailand. **Regional Range:** Vietnam, Cambodia, and Thailand.

91

Long-Headed Lizards Genus *Pseudocalotes*

This genus comprises long-headed, weak-limbed, and shorter-tailed species. The dorsal scales point backward and downward. The dorsal scales also lack uniformity and regularity, a feature distinguishing the genus from other closely related genera. The hind limb reaches to the shoulder or not as far. The tail of the adult male is swollen at the base. Two species occur in Thailand.

KEY TO THE THAI *PSEUDOCALOTES*

1 a	51-61 scale-rows around the body	*P. floweri*
b	65-72 scale-rows around the body	*P. microlepis*

Key modified from Smith (1935).

92 Flower's Long-headed Lizard *Pseudocalotes floweri*

Snout to Vent: 9.8 cm. Tail: 18.0 cm.

Generally light-brown lizard with indistinct dark-brown spots. Nuchal crest has eight erect light-brown spines. Has a flattened forehead with an inverted Y-shaped series of enlarged scales. Tail is banded light and dark brown. Dorsal scales smooth or feebly keeled, larger than the median ventrals. There are 48-56 scales round the body. **Behaviour:** Arboreal. Found on trees and bushes. **Habitat:** Southeastern monsoonal evergreen forest at 1,000 m. **Range:** Recorded only in Khao Soi Dao Wildlife Sanctuary in Chanthaburi province. **Regional Range:** Cambodia and Thailand.

92

93 Boulenger's Long-headed Lizard *Pseudocalotes microlepis*

Snout to Vent: 9.8 cm. Tail: 18.0 cm.

Slender body is light brown with black specks. Black lines radiate from the eyes. The gular pouch of the male is chocolate. Differs from Flower's Long-headed Lizard in the following features: the upper two or three rows point straight backward; the hind limb reaches to the axilla or shoulder; there are 65-72 scales round the body. **Behaviour:** Arboreal. Found on trees and bushes, foraging among the shrubbery and on the ground until dusk. Sometimes encountered sunning itself on wayside bushes. Roosts in streamside vegetation about 1-1.5 m above streams. **Habitat:** Found in montane and sub-montane forests on highest mountains. **Range:** Northern and northeastern (west) Thailand. Recorded in Chiang Mai, Loei and Chaiyaphum provinces. **Regional Range:** India, Myanmar, Vietnam, Laos, and Thailand.

93

Crested Lizards Genus *Calotes*

Members of this genus have comparatively short heads. The scales of the sides of the body point backward and upward; in *Bronchocela* and *Pseudocalotes* these scales point backward and downward. In all members of the genus occurring in Thailand, the tail of the male is swollen at the base. Thailand has 3 species.

KEY TO THE THAI *CALOTES*

1 a	No fold or pit with small scales in front of shoulder; two spines above tympanum; throat and chin with longitudinal dark streaks; colour generally fawn with brownish or blackish marks or bars; 35-52 scales scales around middle of body	*C. versicolor*
b	A fold or pit with small scales in front of shoulder	**2**
2 a	No postorbital spine present; the fold in front of shoulder is oblique, with fine granular scales in the fold; a cream stripe along upper lip extending well onto shoulders; male with 3-5 large reddish-brown or dark-brown spots on sides; 48-52 scales around body	*C. mystaceus*
b	A large postorbital spine present; 40-48 scales around body	*C. emma*

Key modified from Taylor (1963).

94 Changeable Crested Lizard *Calotes versicolor*

Snout to Vent: 9.8 cm. Tail: 11.4 cm.

Colour generally fawn with brownish or blackish marks or bars. Has two separated spines above the tympanum. Nuchal and dorsal crests are continuous. Throat and chin with two longitudinal dark streaks. Absence of fold or pit in front of the shoulder distinguishes from both Moustached and Forest Crested Lizard. Has 35-52 scales around the body. **Behaviour:** Arboreal. Feeds largely on insects and their larvae. Males can be encountered on concrete posts, isolated tree stumps, or trees trunks which serve as vantage points for observing appoaching females or intruding males. Eggs are buried in soft earth. **Habitat:** Found in a wide range of habitats including agricultural lands, gardens, and scrub. **Range:** Widespread and very common throughout Thailand. **Regional Range:** Pakistan, Afghanistan, India, Nepal, Bhutan, Sri Lanka, Bangladesh, Myanmar, China, Vietnam, Laos, Cambodia, Thailand, Peninsular Malaysia, Singapore, and Indonesia (Sumatra).

94

95 Moustached Crested Lizard *Calotes mystaceus*

Snout to Vent: 13.2 cm. Tail: 27.8 cm.

A large, compressed body, with a relatively large head. Dorsum is greyish-brown, turning bright blue or turquoise during the breeding season. Male has 3-5 large reddish-brown or dark-brown spots on sides. Nuchal and dorsal crests are continuous. Head has a diagnostic cream stripe running from the nostrils, through the supralabials, and terminating on the neck. **Behaviour:** Highly arboreal, keeping to shaded habitats. A mature individual occupies a tree trunk, patrolling up and down it as well as around it, chasing away intruders. It chases large insects found on the trunks and branches of trees. The eggs are buried in the soft earth, and are then guarded by the male. **Habitat:** Found in a wide range of forest types, including tree-lined agricultural lands and grasslands. **Range:** Found throughout continental Thailand, southward to Prachuap Khiri Khan Province. **Regional Range:** India (Andamans and Nicobars), Myanmar, Vietnam, Laos, Cambodia, and Thailand.

95

96 Forest Crested Lizard *Calotes emma*

Snout to Vent: 9.8 cm. Tail: 11.4 cm.

The body is brown-coloured with dark-brown dorsal bars. The nuchal and dorsal crests are continuous. Presence of a single postorbital spine distinguishes the species from Moustached Crested Lizard. The breeding male has striking, diagnostic coloration, being salmon-pink on the head with contrasting indigo sides and forelegs. Has a fold or pit on front of shoulder paved with minute black scales. Fourth toe is distinctly longer than third toe. Has 49-65 keeled scales around the body. **Behaviour:** Arboreal as adults, preferring trees and shrubs of primary forests. Young prefer low bushes in more open habitats where they forage early in the morning. **Habitat:** Found in evergreen forests. **Range:** Found throughout the country. **Regional Range:** India, Myanmar, China, Vietnam, Laos, Cambodia, Thailand, and Peninsular Malaysia.

96

Water Dragons Genus *Physignathus*

This genus comprises a single agamid with a more or less compressed body. It has nuchal, dorsal, and caudal crests. It has no distinct gular sac. The tail is cylindrical or more-or-less compressed. Femoral pores are present. A single species occurs in Thailand.

97 Indochinese Water Dragon *Physignathus cocincinus*

Snout to Vent: 19.8 cm. Tail: 53.2 cm.

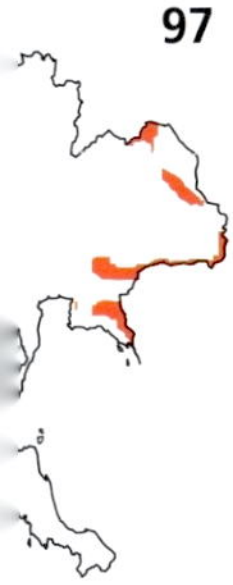

Unmistakable large greenish-olive agamid with a continuous crest running from the neck along the back and down the tail. Above, the head and body are light greenish-olive. Chin is bluish bordered with rose-lavender. Venter is greenish with a wash of greenish-yellow. Strongly compressed tail is grey-green with a series of 12 buff-brown bands. Underside of tail is greyish, more so distally. The tympanum is partly scaled. **Behaviour:** Often found close to forested streams, in trees overhanging the waterways into which it will plunge to escape danger. Forages in the middle of the day for insects,

97a

97b

earthworms, and small rodents. Lays 8-12 eggs in exposed sandy areas on streambeds. It roosts overnight high up in streamside trees. **Habitat:** Inhabits evergreen forest. **Range:** Found in northeastern and southeastern Thailand. Recorded from Sakhon Nakhon, Ubon Ratchathani, Nakhon Ratchasima, Sa Kaeo, Prachinburi, Chanthaburi, Trat, Chonburi, Rayong, Chachaengsao, Nakhon Nayok, and Saraburi provinces. **Regional Range:** China, Vietnam, Laos, Cambodia and Thailand.

Genus *Mantheyus*

This genus comprises an agamid with slender digits at the base, which expand into large fan-shaped expansions. All the digits are clawed. The body is covered with small granular scales which may be intermixed with larger granular tubercles. It lacks preanal and femoral pores. The pupil is vertical. A single species occurs in Thailand.

98 Phu Wua Lizard *Mantheyus phuwuanensis*

Snout to Vent: 8.6 cm. Tail: 18.2 cm.

A large *Mantheyus*, with slender digits, which expand into large fan-shaped expansions. It has three parallel longitudinal folds on each side of the throat. The head and body are dorso-ventrally flattened. Males are coloured dark brown with small greenish blotches and patterns. The black roundish blotches on the brownish to bluish-violet flanks are arranged in oblique rows, proceeding up to the edges of the belly. They are interspersed with pale orange stripes. The throat is yellowish-orange having two rows of black irregular stripes on either side. Lower half of the gular sac is reddish whereas the belly is somewhat bluish-violet intermixed with yellow. In

98

females the greenish blotches and patterns appear yellowish-orange; ventral side is uniformly pale yellow with the dark spots absent. **Behaviour:** Ground dwelling agamid. Shelters in rock crevices. **Habitat:** Found among boulders on slopes of flattened sandstone hills at elevations between 200 and 380 m. **Range:** Northeastern Thailand. Recorded in Phu Wua National Park, Bueng Kan Province. **Regional Range:** Laos and Thailand.

Earless Lizards Genus *Aphaniotis*

This genus comprises a small agamid with a compressed body. The tympanum is hidden. The dorsal scales are small, intermixed with large scales. A single species occurs in Thailand.

99 **Earless Lizard** *Aphaniotis fuscus*

Snout to Vent: 6.3 cm. Tail: 15.3 cm.

A small agamid. Has a very slender body and very long, spindly limbs; these are uniformly dark brown above. Also has a barely distinguishable dorsal crest. Head is olive-brown. The sides of the head are lighter brown with a greenish-blue tinge. The gular pouch is black. Tail has narrow dark and light bands that are not normally symmetrical. **Behaviour:** Arboreal, sometimes being found perched on stems and leaves. The female lays 1-2 eggs. **Habitat:** Found in undisturbed Malayan mixed dipterocarp forest in southernmost provinces. **Range:** Recorded from Pattani and Yala provinces. **Regional Range:** Thailand, Peninsular and Eastern Malaysia, Singapore, and Indonesia (Sumatra).

99

Butterfly Lizards Genus *Leiolepidae*

Members of this genus have a compressed body. They lack a crest. The tympanum is distinct. The tail is long and rounded, and weakly depressed. Femoral pores are present. Three species occur in Thailand.

KEY TO THE THAI *LEIOLEPIS*

1	**a**	Black and orange or yellow bars present along the entire flank of both sexes are clear and complete	**2**
	b	No black and orange or yellow bars are present along the entire length of both sexes	**3**
2	**a**	Numerous large grey or pale-yellow oval spots with black edges forming a reticulum. Longitudinal stripes only appear in the pelvic region. Chin is red in female	*L. belliana ocellata*
	b	Small yellow oval spots and black edges not forming a reticulum. Three or fewer longitidinal stripes on the back	*L. belliana belliana*
3	**a**	Black and orange or yellow bars are present only in male, extending about half of the entire flank. Small dorsal spots with black edges forming a reticulum on the back. Small pale ocelli located within the spots	*L. reevesii*
	b	Two pale yellow continuous longitidinal stripes. Pale ocelli on a dark polygonal network on the back. Light vertical bars are present along the entire flank. Head is light-grey or olive with small pale yellowish spot on the lower eyelid	*L boehmei*

Key modified from Aranyavalai (2003).

100 Common Butterfly Lizard *Leiolepis belliana*

Snout to Vent: 15.6 cm. Tail: 33.7 cm.

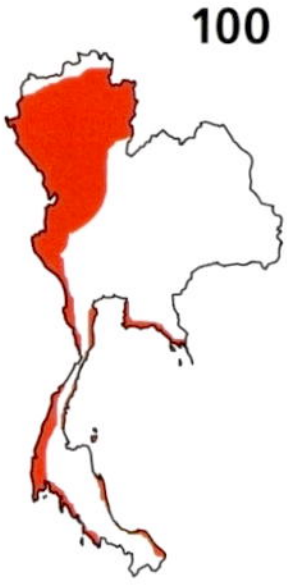

Greyish-olive or blackish above, with pale yellow-edged black spots and usually three or more longitudinal stripes, a median and two laterals. The flanks of the body have bluish-black, with 7-9 orange vertical bars along their entire length. The top of head is dark olive-brown. The female is less brilliantly coloured and the belly is more uniform. **Behaviour:** Terrestrial, living in excavated burrows in loose soils. Feed on insects, particularly during sunny periods. Has a territorial nature, includes extending its ribs—having 'wings' like a butterfly—as it encircles individuals in other burrows. **Habitat:** Found in open forest habitats, favouring dry dipterocarp forest, dry scrub, and coconut groves. **Range:** Northern, western, southeastern, and peninsular Thailand. **Regional Range:** Myanmar, Thailand, Peninsular Malaysia, and Indonesia (Sumatra).

100a

100b

101 Boehme's Butterfly Lizard *Leiolepis boehmei*

Snout to Vent: 11.0 cm. Tail: 24.5 cm.

Smaller in size than other butterfly lizards. This species lacks the characteristic contrasting orange and black vertical flank marking, instead having indistinct light oblique stripes. The dorsum is blackish-olive with two pale-grey lateral stripes, each 6-7 body scales wide. Between these dorso-lateral stripes are a series of 6-7 grey oval spots aligned in indistinct transverse rows. The head is olive-coloured, with darker areas on the sides and a small yellowish dot on the lower eyelids. The chin, throat, and breast are grey with white vertical bars. **Behaviour:** Terrestrial. Lives in burrows. **Habitat:** Found in open forest habitat in coastal areas. **Range:** Recorded in Nakhon Si Thammarat and Songkhla provinces. **Regional Range:** Endemic to Thailand.

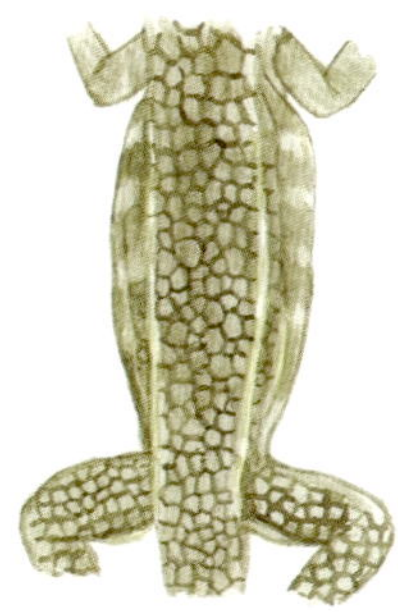

101

102 Eastern Butterfly Lizard *Leiolepis reevesii*

Snout to Vent: 15.1 cm. Tail: 34.0 cm.

Distinguished from Common Butterfly Lizard by having diffuse, buffy-brown spot markings on the dark brown dorsal surface giving a fish-net pattern. The flanking pattern, consisting of a couple of contrasting orange vertical stripes on black flanks is much reduced, being confined to an area just behind the forelimbs; by comparison, the rich-orange horizontal flank markings are more extensive. The dark-brown dorsal surfaces of both hind-limbs and the tail are strongly spotted with whitish-yellowish spots. **Behaviour:** Diurnal and terrestrial, living in excavated burrows. Territorial, with threat displays involving stretching the ribs to show the brightly coloured flanks. Diet includes grasshoppers, butterflies, and other insects. **Habitat:** Found in open forests and arid habitats on the Khorat Plateau. **Range:** Northeastern Thailand. Recorded in Chaiyaphum, Khon Kaen, Amnart Charoen, Sisaket, and Nakhon Ratchasima provinces. **Regional Range:** China, Vietnam, Laos, Cambodia, and Thailand.

102

Long-tailed Lizard Family Lacertidae

Long-Tailed Lizards Genus *Takydromus*

This genus have cylindrical tails, which are either long or very long. The lower eyelid is scaly. The dorsal has large, strongly keeled plates which form continuous lines. The digits are more-or-less cylindrical. One species occur in Thailand.

103 Long-tailed Lizard *Takydromus sexlineatus*

Snout to Vent: 6.5 cm. Tail: 36.0 cm.

Easily recognized lizard with an extremely long tail, which is nearly six times as long as the head-body length. Dark-olive head, greenish-olive dorsal surface and dark horizontal stripe passing from the snout, through the eye and above the tympanum, down each side of the body provides good

103

camouflage. Scale-rows of neck and upper half of body strongly keeled. **Behaviour:** Diurnal, foraging through loosely woven grass, particularly long grass stalks, and bamboo. It is capable of moving at considerable speed. Lays 2-4 eggs. **Habitat:** Found throughout Thailand. Recorded in Mae Hong Son, Chiang Mai, Bangkok, Nakhon Si Thammarat, and Pattani provinces. **Regional Range:** India, Myanmar, China, Vietnam, Laos, Cambodia, Thailand, Peninsular and Eastern Malaysia, Singapore, and Indonesia (Sumatra and Java).

Skinks Family Scincidae

The Skinks are almost smooth, shiny lizards, the vast majority of which are very active, alert, and terrestrial. A number have become arboreal or at least semi-arboreal. In a few genera—namely the Legless Skinks *Davewakeum* and *Isopachys*—the species have lost parts of the limbs and the animals are becoming subterranean. Most species live amongst the leaf litter and loose soil, foraging by day and taking shelter at night in forest debris, as well as under logs and rocks. Some species, notably the Stream Skinks of Africa, South-east Asia, and Australia, are semi-aquatic. Insects and other arthropods are the main prey, though some larger skinks can eat eggs and young birds. The family comprises a large number of genera: the Sun Skinks *Eutropis* (5 species); the Blue-tailed Skink *Plestiodon*; the Striped Tree Skinks *Lipinia* (2 species); the Tree Skink *Dasia* (1 species); the Stream Skinks *Tropidophorus* (8 species), the Leaf-litter Skinks *Scincella* (4 species); the Forest Skinks *Sphenomorphus* (11 species); the Supple Skinks *Lygosoma* (8 species); the Larut Skink *Leptoseps* (1 species); and the Legless Lizards comprising *Isopachys* (4 species) and *Davewakeum* (1 species). More than 1,300 species are known worldwide while 48 species occur in Thailand.

KEY TO THAI GENERA OF THE SCINCIDAE

1 a	Palatine bones separated on the midline of the palate; nostril pierced in a single nasal, supranasals present; a paired frontoparietal	*Plestiodon*
b	Palatine bones meet on the midline of the palate	**2**
2 a	Pterygoid bones separated; palatal notch extending forward to level of middle of eyes; supranasals present; legs pentadactyl, well-developed	*Eutropis*
b	Pterygoid bones usually meeting on palate; palatal notch not reaching to middle of eyes	**3**
3 a	Limbs absent	**4**
b	Limbs present	**5**
4 a	Fourth supralabial below middle of eye; dorsum uniform or with a faint dark line on each scale-row	*Davewakeum*
b	Second or third supralabial below middle of eye; dorsum distinctly patterned	*Isopachys*
5 a	Limbs with less than five digits	*Leptoseps*
b	Limbs with five fingers or toes	**6**
6 a	Lower eyelid scaly, lacking a more or less transparent or translucent disc; occasionally with a small opaque disc	**7**
b	Lower eyelid with a more or less transparent or translucent disc	**13**
7 a	Supranasal present (sometimes fused partially with the nasals)	**8**
b	Supranasal absent	**11**
8 a	Ear opening present	**9**
b	Ear opening absent, or else covered by overlapping scales	*Dasia*
9 a	Supranasals reduced, separated by the rostrofrontonasal suture	*Lygosoma* (a)
b	Supranasals enlarged, forming a median suture	**10**
10a	A dark flank-stripe present	*Riopa*
b	No dark flank stripe	*Lygosoma* (b)
11a	Tympanum sunk (if visible at all)	**12**
b	Tympanum superficial	*Tropidophorus*
12a	Frontoparietal united	*Lygosoma*-(c)
b	Frontoparietal paired	*Sphenomorphus*
13a	Ear concealed	*Lipinia*
b	Ear opening present	*Scincell* a

Key modified from Dring (unpublished).

Sun Skinks Genus *Eutropis*

Members of this genus are ground-dwelling and have sturdy bodies and strong well-developed limbs for movements in the leaf litter and loose soil. They have well-developed eyelids. The ear has a deeply sunken tympanum. Five species are found in Thailand.

KEY TO THE THAI *EUTROPIS*

1	**a**	No postnasal	**2**
	b	Postnasal present	**3**
2	**a**	Lower eyelid with transparent disc; adult SVL below 8 cm	*E. novemcarinata*
	b	Lower eyelid with several scales instead of a disc; adult SVL below 8 cm	*E. macularia*
3	**a**	Dorsum with 5-7 greenish-white longitudinal lines or olive unicoloured; 14-28 scale-rows around middle of body; adult SVL below 8 cm	*E. rugifera*
	b	Dorsum lacking 5-7 greenish-white longitudinal lines; number of keels variable; not olive unicolour	**4**
4	**a**	Scales finely striate, with two or three feeble keels; 26-30 scale-rows around middle of body; adult SVL above 10 cm	*E. longicaudata*
	b	Three to five keels on dorsal scales; 30-34 scale-rows around middle of body; adult SVL above 10 cm	*E. multifasciata*

Key prepared by Boehme (per comms).

104 Many-lined Sun Skink *Eutropis multifasciata*

Snout to Vent: 12.5 cm. Tail:18.0 cm.

A large, sturdy sun skink. Bronze or olive grey above, with narrow black lines bordering scale-rows giving a scaly pattern. The usually characteristic broad brown lateral stripe—typical for the genus—which also runs from behind the eye along the sides of the body to the base of the tail is lighter brown or grey brown on the lower sides, and consequently much less contrasting. A yellowish-olive or light-olive lateral stripe runs from the snout, above the eye and down the length of the body onto the tail. The flanks have a series of white spots. The dorsal scales have 3-5 keels. Venter is greyish or greenish-white. Labials are greenish-white. It has 30-34 scale-rows around the middle of the body. **Behaviour:** Terrestrial, being found in leaf litter on the forest floor. Mainly diurnal, though also active at dusk. **Habitat:** Lowlands. Common around human settlements. **Range:** Found throughout Thailand, being recorded from virtually every province. **Regional Range:** India, Myanmar, China, Vietnam, Laos, Cambodia, Thailand, Peninsular and eastern Malaysia, Singapore, Brunei, Indonesia, the Philippines (Palawan), East Timor, and Papua New Guinea.

104

105 Long-tailed Sun Skink *Eutropis longicaudata*

Snout to Vent: 11.7 cm. Tail: 20.0 cm.

A largish sun skink with a diagnostic long tail which is almost twice as long as the body. Head and body are uniformly brown or brownish-olive above, with a broad dark brown or black lateral stripe starts behind the eye, running along the sides to the base of the tail. The body scales are finely striate, with 2-3 poorly defined keels. The labials are greenish with some black markings. There are 26-30 scale-rows around the middle of the body. **Behaviour:** Mainly terrestrial, being found in leaf litter on the forest floor, though also exhibits arboreal habits. **Habitat:** Found at higher altitudes. **Range:** Widespread throughout Thailand. **Regional Range:** China, Vietnam, Laos, Cambodia, Thailand and Peninsular Malaysia.

105

106 Keeled Sun Skink *Eutropis novemcarinata*

Snout to Vent: 7.1 cm. Tail: 7.4 cm.

A small sun skink. Dark brown head and back with a broad, with a dark brown stripe running from behind the eye along the side of the body onto the tail. Beneath this stripe are numerous bluish-white speckles. On the dark brown back are two series of indiscrete blackish markings start at the shoulders, becoming larger and more clearly defined along the back to the base of tail. Dorsal and lateral scales have 7-11 sharp keels. Lower eyelid has an undivided transparent disc. The venter is greenish-white. It

106

has 32-34 scale-rows around the body. **Behaviour:** Diurnal and terrestrial, being found in leaf litter on the forest floor. **Habitat:** Occurs in peninsular monsoonal evergreen forest and Malayan mixed dipterocarp forest. **Range:** Found in peninsular Thailand. Recorded from Nakhon Si Thammarat, Trang, Pattani, Narathiwat, and Yala provinces. **Regional Range:** Myanmar, Thailand, and Peninsular Malaysia.

---

107 Nicobar Sun Skink *Eutropis rugifera*

Snout to Vent: 5.4 cm. Tail: 11.4 cm.

One of the smallest sun skinks. Head and back are dark olive, with 5-7 greenish-white longitudinal lines running along the back. Sides of head are brownish. The venter is greenish-white. Has 24-28 scale-rows around the middle of the body. **Behaviour:** Diurnal and terrestrial, being found in leaf litter on the forest floor. Feeds on arthropods. **Habitat:** Inhabits Malayan mixed dipterocarp forest. **Range:** Extreme southern provinces of peninsular Thailand. Recorded from Hala Bala Wildlife Sanctuary, Narathiwat Province and Bannang Sata, Yala Province. **Regional Range:** India (Nicobars), Thailand, Peninsular and Eastern Malaysia, and Indonesia (Sumatra and Java).

107

---

108 Variable Sun Skink *Eutropis macularia*

Snout to Vent: 5.8 cm. Tail: 9.0 cm.

A small sun skink. This species shows considerable variation in coloration, and is recognized by three sub-species. *Mabuya m. postnalis* is olive-brown or brown above with a diffuse lateral band on the sides. *Mabuya m. malcolmi* is generally olive with a slightly darker head. It has a dorsolateral stripe on the side of the head and neck that extends on the anterior part of the body. *Mabuya m. quadrifasciata,* which is confined to the Khorat Plateau, is olive-brown above with four dark dotted lines on the back; the scales on throat and chin have brown borders. Ventral surface is ultramarine. All sub-species have five, seven, or nine keels present at various points on the body and all sub-species have specialized scales that form "chigger-mite" refuges on the lower legs above the ankles. **Behaviour:** Nocturnal, foraging at night close to waterways. **Habitat:** Found in mixed deciduous and dry dipterocarp forest. Also found in secondary forest and plantations. **Range:** Found throughout continental Thailand and northern peninsular Thailand. **Regional Range:** Pakistan, India, Sri Lanka, Bangladesh, Myanmar, Vietnam, Laos, Cambodia, and Thailand.

108

Tailed Skinks Genus *Plestiodon*

Members of this genus have well-developed, pentadactyl limbs. They have a distinct, deeply sunk, tympanum. The eyelids are well-developed. The nostril is pierced in the nasal. The lateral teeth are conical or with spheroidal crowns. They guard their eggs. Worldwide, some 35 species are known; in Thailand a single species is present.

109 **Blue-tailed Skink** *Plestiodon quadrilineatus*

Snout to Vent: 7.7 cm. Tail: 11.5 cm.

Smallish-sized skink. Dark-brown or olive-brown body, with contrasting silvery stripes running from the snout along the sides to near the end of the tail. A second lateral line begins on the lips and extends along the flanks to the base of the tail. The posterior half or third of the tail is blue. The throat is whitish. The ventral surface is creamy brown. The young are bluish-black above, with the same longitudinal stripes. **Behaviour:** Diurnal and terrestrial. **Habitat:** Found at high elevations. **Range:** Northern and central Thailand. Recorded in Chiang Mai (Doi Chiang Dao) and Saraburi provinces. **Regional Range:** China, Vietnam, Cambodia, and Thailand.

109

Striped Skinks Genus *Lipinia*

This genus comprises small striped skinks. The eyelid is well-developed. The ear opening is distinct. Two species occur in Thailand.

KEY TO THE THAI *LIPINIA*

1 a Distance between end of snout and forelimb about the same length as the distance between the axilla and groin; striping down length of body *L. vittigera*

b Distance between end of snout and forelimb nearly twice the distance between the axilla and groin; uniform body coloration lacks striping *L. surda*

Key modified from Dring (unpublished).

110 Common Striped Skink *Lipinia vittigera*

Snout to Vent: 3.6 cm. Tail: 4.0 cm.

A diminutive, delicate skink. Diagnosed by a strong striping pattern running the length of the body. This striping pattern comprises a broad central greenish-white, yellow or cream-coloured stripe extending from the tip of the snout until the tip of the tail; on either side are two black or brown bands extending to the base of the tail. Beneath these are two additional lateral yellow or cream stripes. Ventral surface is greenish-white. The distance between the end of strongly pointed snout and forelimb is about the same length as the distance between the axilla and groin. **Behaviour:** Diurnal and arboreal, being encountered on tree trunks, though possibly living high into the canopy. Diet consists of small insects. **Habitat:** Found in a wide range of forest types, including lower montane forest, dry evergreen forest, mixed deciduous forest, and southeastern monsoonal evergreen forest. Occurs from level plains up to at least 1,500 m. **Range:** Found throughout Thailand as far south as the Isthmus of Kra. **Regional Range:** Myanmar, Vietnam, Laos, Cambodia, Thailand, and Peninsular Malaysia.

110

111 Selangor Striped Skink *Lipinia surda*

Snout to Vent: 5.0 cm. Tail: 6.7 cm.

Uniform dark brown with a reddish tinge above and greyish underneath. The snout is pointed. Lower eyelid has an undivided semi-transparent disc. Distance between the end of the snout and forelimb is nearly twice the distance between the axilla and groin. The short limbs are well-developed; they are widely separated when pressed against the body. There are 20 scale-rows around the body. **Behaviour:** Diurnal and arboreal. **Habitat:** Found in roots of epiphytic plants and in dead fibres of palm trees. Inhabits Malayan mixed dipterocarp forest. **Range:** Recorded from Yala Province in extreme southern Thailand. **Regional Range:** Thailand and Peninsular Malaysia.

111

Tree Skinks Genus *Dasia*

Members of this genus are arboreal, and their digits show the earliest stage of the evolution of the adhesive digital pad. They have well-developed limbs which are pentadactyl. The eyelids are well-developed. The ear opening is small, with the tympanum deeply sunken. One representative of the genus is found in Thailand.

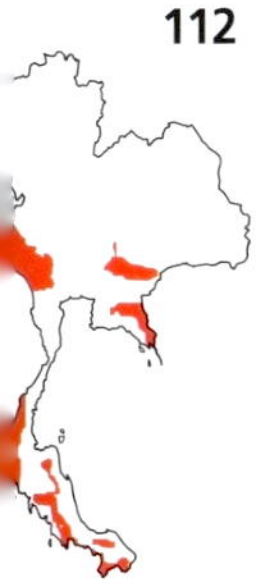

112 Olivaceous Tree Skink *Dasia olivacea*

Snout to Vent: 11.5 cm. Tail: 15.7 cm.

A moderately large skink. Greenish-brown above and on the sides sometimes with black spots arranged in transverse series; many of these spots have a white centre. The venter is pale blue, green, or yellowish. The tail is olivaceous or blackish. The young are black above, with narrow silvery or yellowish transverse bars on the neck and body. There are 30 scale-rows around the mid-body. **Behaviour:** Diurnal and arboreal. Diet consists of arthropods, including ants and beetles. **Habitat:** Occurs in semi-evergreen forest, monsoonal evergreen forest and Malayan mixed dipterocarp forest, as well as mangrove and coconut plantations. **Range:** Found in western (southern), southeastern, and peninsular Thailand. Recorded in Nakhon Ratchasima, Sa Kaeo, Chanthaburi, Kanchanaburi, Surat Thani, Phuket, Phang-nga, Trang, Pattani, Satun, and Yala provinces. **Regional Range:** India (Andamans and Nicobars), Myanmar, Vietnam, Cambodia, Thailand, Peninsular and Eastern Malaysia, Singapore, and Indonesia (Sumatra and Java).

112

Stream Skinks Genus *Tropidophorus*

A genus of mainly semi-aquatic lizards living in the vicinity of rocky streams, hiding under stones or debris during the daytime. Some members of the genus live in rock crevices. At dusk they become active in their search for food. They are not remarkably agile. In their general body plan and plain coloration they are reasonably similar, though the species found in crevices have a depressed head and body. The limbs are well-developed and pentadactyl. The digits are cylindrical. The eyelids are well-developed. Eight species are found in Thailand.

KEY TO THE THAI *TROPHIDOPHORUS*

1 a	Tail spiny	*T. hangnam*
b	Not as above	**2**
2 a	Three large preanal scales; dorsal scales strongly keeled in juvenile and adult	*T. microlepis*
b	Two large preanal scales	**3**

3 a	Body relatively depressed	**4**
b	Body not depressed	**5**
4 a	Frontonasals separated; mid-body scales 25	*T. latisculatus*
b	Frontonasals in contact; mid-body scales 34	*T. matsuii*
5 a	Head shields smooth; dorsal scales in juveniles keeled but in adult adult more or less smooth, at least on foreparts	**6**
b	Head shields rugose or finely scultured; dorsal scales in juvenile and adult strongly keeled	**7**
6 a	Frontonasal undivided.	*T. berdmorei*
b	Frontonasal divided along midline usually	*T. laotus*
7 a	Frontonasal and frontal undivided; flank scale-rows directed straight backwards	*T. robinsoni*
b	Frontonasal divided along midline; anterior part of frontal broken up in addition to a small scale between prefrontals; some flank scale-rows directed obliquely	*T. thai*

Key modified from Smith (1935).

113 Spiny Stream Skink *Tropidophorus hangnam*

Snout to Vent: 7.8 cm. Tail: 6.5 cm.

Dorsal part of head reddish-brown; dorsal and lateral part of neck, body, limbs and tail blackish-brown with transverse interrupted creamy white bands; ventral part of hands and feet grey. The tail is shorter than the head and body. Head scales smooth; dorsal weakly keeled, lateral body keeled, and upper, anterior and posterior parts of limbs keeled, and tail strongly keeled; lateral scale-rows directed obliquely; two large preanal scales, and sub-caudals paired. **Behaviour:** Mainly aquatic, found in amongst rocks. Feed upon insects, small crustaceans and worms. **Habitat:** Found in small, rocky, forest streams in sub-montane forests. **Range:** Found in northeastern Thailand. Recorded from Phu Khieo Wildlife Sanctuary, Chaiyaphum Province. **Regional Range:** Endemic for Thailand.

113

114 Berdmore's Stream Skink *Tropidophorus berdmorei*

Snout to Vent: 8.5 cm. Tail: 10.4 cm.

One of the largest stream skinks. The smooth body is chocolate brown on the back and tail, with transverse pale brown bars surrounded by black areas. The sides of the body and tail have numerous bluish-white flecks. The venter is pale yellow, though may be salmon pink, extending to underside of limbs. The upper head shields are smooth; in many congeneric species the upper head shields are rugose or striated. The dorsal scales are smooth or obtusely keeled. The limbs are small. **Behaviour:** Diurnal. Mainly aquatic, found in amongst rocks. Feed upon insects, small crustaceans and worms. Ovoviviparous; the females carrying fertilized eggs within their bodies until the embryos are fully developed. **Habitat:** Found in small, rocky, forest streams in sub-montane forests. **Range:** Found in northern, northeastern (west), and western Thailand. **Regional Range:** Myanmar, China, and Thailand

114

115 Laotian Stream Skink *Tropidophorus laotus*

Snout to Vent: 7.5 cm. Tail: 10.8 cm.

Closely resembles Berdmore's Stream Skink, with body coloration virtually identical. Main identification feature is the fronto-nasal, which is divided in the species; in Berdmore's Stream Skink the fronto-nasal is entire. The body is dark brown on the back and tail, with contrasting paler brown bars surrounded by black areas. The sides of the body and tail have numerous bluish-white flecks. The scales following the vent are strongly widened. It has three pairs of nuchals and has 36 scale-rows. **Behaviour:** The natural history of this species is largely unknown. Found in small, rocky, forest streams in sub-montane forests. **Habitat:** Sub-montane and montane forest of high mountains. **Range:** Phu Luang Wildlife Sanctuary and Phu Kradung National Park, Loei Province and Phu Wua National Park, Bueng Kan Province. **Regional Range:** Laos and Thailand.

115

116 Robinson's Stream Skink *Tropidophorus robinsoni*

Snout to Vent: 6.4 cm. Tail: 9.1 cm.

Generally more slender than Berdmore's Stream Skink, with a dark brown or blackish head, body, and tail. It has all-light brown bands with black edges along the length of the upper body; two cross-bands on the neck; 3 transverse bands on the body and 13 indistinct transverse bands on the tail. The upper head shields are rugose or feebly striated. The throat and venter are ivory yellow. Underside of tail is ivory yellow, thickly spotted with dark brown. Limbs are noticeably shorter. There are 30-32 scale-rows around the mid-body. **Behaviour:** Unknown, as no description was provided when the specimen was taken. **Habitat:** Found in peninsular monsoonal evergreen forest. **Range:** Recorded in Tasan, Chumphon Province in upper peninsular Thailand. **Regional Range:** Myanmar and Thailand.

116

117 Thai Stream Skink *Tropidophorus thai*

Snout to Vent: 7.4 cm. Tail: 7.9 cm.

Light brown above, with numerous pale brown or yellow dark-edged markings on the back; these markings are sometimes V-shaped. It has two transverse markings on the neck; nine transverse bands on the body and nine transverse bands on the tail. The throat and venter are brownish-white. The tail is mottled with brown. There are 38 scales around the middle of the body. **Behaviour:** Diurnal and semi-aquatic. **Habitat:** Found along streams in upper montane forest. **Range:** Restricted to extreme northwestern Thailand, in Mae Hong Son and Chiang Mai provinces. **Regional Range:** Endemic to Thailand.

117

118 Cambodian Stream Skink *Tropidophorus microlepis*

Snout to Vent: 8.3 cm. Tail: 14.2 cm.

Brown or reddish-brown above, with almost distinct, black-edged, pale spots. The dorsum is also indistinctly marked with transverse, darker brown bands. Lips and flanks with pale spots. Limbs and digits are dark banded. Brownish-white below. It has three preanal shields. There are 28-32 scales around the mid-body. **Behaviour:** Semi-aquatic, foraging close to waterways. Nocturnal, hiding by day beneath damp vegetation or under stones. **Habitat:** Found in southeastern monsoonal evergreen forest. **Range:** Restricted to southeastern Thailand. Recorded in Khao Sabab and Khao Soi Dao Wildlife Sanctuary, Chanthaburi Province, and Pang Sida National Park, Sa Kaeo Province. **Regional Range:** Vietnam, Cambodia and Thailand.

118

119 Matsui's Stream Skink *Tropidophorus matsuii*

Snout to Vent: 9.4 cm. Tail: 11.3 cm.

Stream skink with a moderately depressed head, body, and tail. The dorsal surfaces and sides of the head, body and tail are dark brown. It has pale-brown transverse bands along the length of is body, with three bands on the neck; 9 bands on the body and 23 bands on the tail. It has several irregular pale-brown spots on the sides of the body. The throat, venter, and underside of the tail are ivory-yellow; the tail has dark flecks. **Behaviour:** Little known. **Habitat:** Rock crevices in evergreen forest. **Range:** Recorded in Phu Pa Namtip in Roi Et Province, in northeastern Thailand. **Regional Range:** Endemic to Thailand.

119

120 Broad-scaled Stream Skink *Tropidophorus latiscutatus*

Snout to Vent: 9.1 cm. Tail: 9.6 cm.

A stream skink with a moderately depressed head, body, and tail. The dorsal surfaces and sides of the head, body, and tail are dark brown. It has irregularly shaped, pale brown, transverse bands on the body; two bands on the neck; nine bands on the body and seven bands on the tail; all pale brown. It has several pale brown spots on the supralabials and infralabials. The throat and venter are ivory-yellow. **Behaviour:** Thought to be only very partially aquatic, living in moist crevices close to marshy habitat. Probably nocturnal. **Habitat:** Found in moist crevices in sandstone rocks

120

and outcrops, in dry dipterocarp forest. **Range:** Known in Phu Wua Wildlife Sanctuary in Bueng Kan Province, northeastern Thailand. **Regional Range:** Endemic to Thailand.

Leaf-litter Skinks Genus *Scincella*

Members of this genus are small to medium-sized lizards, with sturdy bodies and strong well-developed limbs for movements on the ground. The movable lower eyelid has a transparent window. All body scales are smooth. In all species the limbs are present and pentadactyl. The dorsum is usually brown to bronze brown. This genus is usually characterized by a dark-brown lateral band starting from the nostrils, running backward through the eye and the ear, to the hind legs. Four species are found in Thailand.

KEY TO THE THAI *SCINCELLA*

1	**a**	Dorsal and lateral scales equal or dorsals a little larger; no enlarged nuchals; ear opening larger than palpebral disc, without projecting lobules	**2**
	b	Dorsal scales distinctly larger than laterals, four or six across the back; 3-6 pairs of nuchals; 28-32 scales around body	*S. doriae*
2	**a**	The adpressed limbs nearly meet or overlap	**3**
	b	The adpressed limbs widely separated; distance between end of snout and forelimb twice the distance between the axilla and the groin; third and fourth toes equal	*S. punctatolineata*
3	**a**	28-32 scales round body; back with small black spots	*S. reevesii*
	b	34-38 scales round body; same colour	*S. melanosticta*

Key modified from Taylor (1963); Ouboter (1986).

121 Doria's Leaf-litter Skink *Scincella doriae*

Snout to Vent: 5.8 cm. Tail: 10.1 cm.

A large robust, leaf-litter skink, with a caramel brown body. A dark brown lateral band runs from the nostril, through the eye and above the ear, to behind the hind legs; this band is broken up by whitish spots. The margins of the labials are edged with brown. Ventral surface and tail is

121

yellowish-white. **Behaviour:** Diurnal and terrestrial. Inhabits debris on the forest floor. **Habitat:** Found in upper montane forests. **Range:** Presently recorded in Chiang Mai Province, northern Thailand, and Phu Luang Wildlife Sanctuary, Loei Province, northeastern Thailand. **Regional Range:** Myanmar, China, and Thailand.

122 Black-spotted Leaf-litter Skink *Scincella melanosticta*

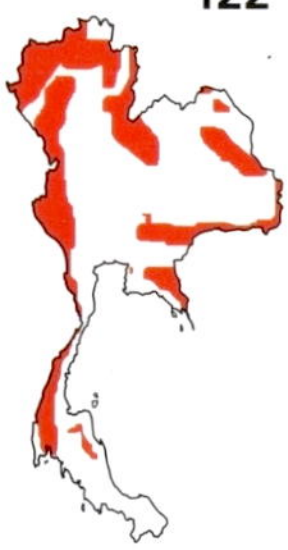

Snout to Vent: 5.0 cm. Tail: 8.5 cm.

A medium-sized, rather robust leaf-litter skink with a broad head and rather long limbs. Dorsum is olive brown or bronze brown, marked with brown or black spots. A dark brown lateral band is broken up by creamish or whitish spots. Sides and upper parts of limbs are heavily marked with small, dark brown spots. Has 32-38 scales around its mid-body; by comparison, Reeve's Leaf-litter Skink has 28 to 32 scales around its mid-body. **Behaviour:** Diurnal and terrestrial, inhabiting the leaf litter and other forest floor debris. **Habitat:** Found in a wide range of evergreen forest types. **Range:** Widely distributed throughout the country as far south as Khao Luang National Park, in peninsular Thailand. **Regional Range:** Myanmar, Vietnam, Cambodia and Thailand.

122

123 Reeve's Leaf-litter Skink *Scincella reevesii*

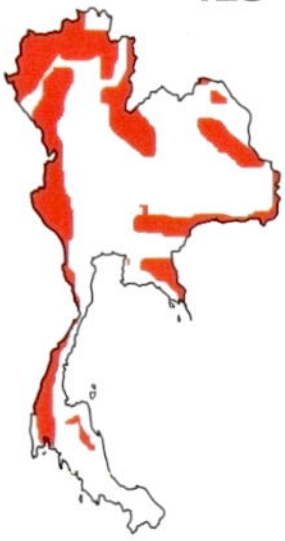

Snout to Vent: 4.1 cm. Tail: 7.2 cm.

A medium-sized, somewhat elongated, leaf-litter skink. Dorsum is bronze brown, with black spots concentrated in the vertebral region. Dark brown lateral band is broken up by cream or whitish spots. Sides and upper parts of limbs are heavily marked with small, dark-brown spots. Belly and tail is yellowish-white. It has 28-32 scales around mid-body. **Behaviour:** Diurnal and terrestrial, living in the leaf litter and other forest-floor debris. **Habitat:** Found in a wide range of forest types. **Range:** Widely distributed in forested areas throughout the country as far south as Nakhon Si Thammarat, in peninsular Thailand. **Regional Range:** Myanmar, China, Vietnam, Laos, Cambodia, and Thailand.

123

124 Western Leaf-litter Skink *Scincella punctatolineata*

Snout to Vent: 4.0 cm. Tail: 6.4 cm.

Light brown above, with small dark brown spots forming longitudinal streaks. It has a dark dorso-lateral stripe, which is two scales wide. The flanks are spotted with brown. Venter is creamish. The adpressed limbs are widely separated; in Black-spotted Leaf-litter Skink and Reeve's Leaf-litter Skink the adpressed limbs nearly meet or overlap. The distance between the end of the snout and the forelimb is twice the distance between the axilla and the groin. It has 24 or 26 scales around mid-body. **Behaviour:** Diurnal and terrestrial. Lives in the leaf litter. Vulnerable to increasing frequency of forest fires. **Habitat:** Presently only known in mixed deciduous forest. **Range:** Confined to Sai Yok National Park and Erawan National Park in Kanchanaburi Province, western Thailand. **Regional Range:** Myanmar and Thailand.

124

Forest Skinks Genus *Sphenomorphus*

Members of this genus are smooth-scaled terrestrial lizards. The limbs are well-developed and pentadactyl. The lower eyelid is scaly. Eleven species are found in Thailand.

KEY TO THE THAI *SPHENOMORPHUS*

1 a	Adpressed limbs barely touch, or fail to touch when adpressed, diminutive skinks (SVL < 40 mm)	2
b	Adpressed limbs overlap at least length of hands, larger species (SVL > 40 mm)	3

2 a	Sub-caudals enlarged, one pair of nuchals; 30 mid-body scale-rows	*S. mimicus*
b	Sub-caudals not enlarged; no nuchals; 34 mid-body scale-rows	*S. grandisonae*
3 a	Mid-body scale-rows not more than 30; 2 or 3 pairs of nuchals	**4**
b	Mid-body scale-rows 30 or more; one pair of nuchal, or none	**5**
4 a	Mid-body scale-rows 28; three pairs of nuchals; prefrontal in contact	*S. praesignis*
b	Mid-body scale-rows 22-24; two or three nuchals; prefrontals not in contact	*S. stellatus*
5 a	Six supralabials, 4th and 5th below eye	*S. butleri*
b	Seven or eight supralabials, 5th and 6th below eye	**6**
6 a	Scale-rows around middle of body number 30	*S. helenae*
b	Scale-rows around middle of body number more than 30	**7**
7 a	Rostral flat or concave; 42-44 scale-rows round middle of body	*S. maculatus*
b	Rostral not flat or concave	**8**
8 a	Dorsum with small somewhat uniform spots arranged in longitudinal and transverse rows; scale-rows 34-36; dorsal scales a little wider than the laterals	*S. tarsus*
b	Not so marked; some or most dorsal scales wider than laterals	**9**
9 a	Dorsolateral line broken into a series of dark-brown spots, each of which may have two prongs on lower sides, and separated from each other by fawn or buff spots; mid-body scale-rows 34	*S. scotophilus*
b	Dorsolateral line normally continuous; large species	**10**
10a	Lateral dark stripe variously flecked and spotted with white, some irregular rows of dark spots on dorsum; scale-rows 38; no nuchals	*S. lineopunctatus*
b	Lateral stripe black without white flecks or spots; scale-rows 34-36; a pair of nuchals	*S. indicus*

Key modified from Taylor (1963).

125 Mimic Forest Skink *Sphenomorphus mimicus*

Snout to Vent: 3.6 cm. Tail: 5.8 cm.

A small slender-bodied skink. Above variegated fawn with clouding or small indistinct brown spots, particularly on the shoulders. Both the supralabials and infralabials have dark-brown spots. The sides of the head are reticulated with brown. The side of the neck and sides with the flecks of brown. The arms and the legs are strongly spotted with brown and fawn. The chin, venter, underside of limbs, and sub-caudal region are cream white. Prefrontals in contact; no supranasals, three supraoculars touch frontal; no post nasal; two frontoparietals; parietals enclose interparietal; one pair of nuchals; seven supralabials, 5th and 6th enlarged; six infralabials, first small; one pair large preanals. Thirty scale-rows about body at middle;

ventral scales larger than dorsals; sub-caudal scales enlarged. Lower eyelid scaly. **Behaviour:** Diurnal and terrestrial. Found in leaf litter. **Habitat:** Found in dry evergreen forest. **Range:** Found in northeastern and central Thailand, Recorded in Saraburi, Nakhon Ratchasima, and Loei provinces. **Regional Range:** Endemic to Thailand.

126 Grandison's Forest Skink *Sphenomorphus grandisonae*

Snout to Vent: 3.0 cm. Tail: 2.1 cm (tip broken).

A very small slender-bodied forest skink. Above light brown with some minute flecking on dorsal scales but no series of spots. The head with supraorbital sutures darkened, as are many other hand sutures. The lips have dark bars on the labial sutures, separated by cream spots. It has a lighter dorsolateral line on the side of the neck and shoulder which does not continue along the side except as a very indefinite line of scarcely discernable darker flecking with some lighter dots. There are dark and light indistinct marks on the limbs and tail. The chin, venter, and tunderside of tail are white or cream; tail. There are 34 scale-rows around the body. Detailed key features comprise the following: six supralabials; five infralabials; median suboculars larger than pre- or post-suboculars, the whole series continuous; temporals 2+3+4, no nuchals; dorsal scales smaller than ventrals; sub-caudals smaller than the lateral caudal scale. **Behaviour and Habitat:** Nothing known of its natural history. **Range:** Found in northern Thailand, being recorded from Ban Tong Pheung, Uttaradit Province. **Regional Range:** Endemic to Thailand.

127 Common Forest Skink *Sphenomorphus maculatus*

Snout to Vent: 5.3 cm. Tail: 9.5 cm.

A small forest skink. Body is brownish-olive or bronzy above, becoming lighter colour along the tail. It has two series—or rarely one vertebral row—of small black dots on the two median scale-rows. It also has a dark brown or black lateral band, spotted with white and usually edged with white below from the snout to the tail. The tympanum is on the surface and not sunk. The top of the head has dark areas above the orbits. The snout is short and obtuse. The rostral is flat or concave. There are 42-44 scale-rows

127

round the middle of the body. It has 128 widened sub-caudals. The limbs are well-developed. **Behaviour:** Diurnal and terrestrial. Diet consists of insects and spiders. **Habitat:** Found close to streams. Inhabits primary evergreen forests. **Range:** Continental Thailand extending southward to the Isthmus of Kra, in Ranong/Chumphon provinces. **Regional Range:** India, Myanmar, China, Vietnam, Laos, Cambodia, and Thailand.

128 Pale Forest Skink *Sphenomorphus tersus*

Snout to Vent: 9.0 cm. Tail: 17.0 cm.

One of the larger forest skinks. The back is dark brown with numerous spots in longitudinal and transverse rows, which become less distinct posteriorly. This species lacks a dark lateral stripe passing through the head and along the flanks. Flanks are lighter, either pale brown or cream. Above each eye is a black area. It has distinct ear openings. Venter is cream-coloured. It has 34-36 scale-rows around the mid-body. **Behaviour:** Diurnal and terrestrial. Found in leaf litter. **Habitat:** Found in lowlands in peninsular monsoonal evergreen forest. **Range:** Found throughout peninsular Thailand northward to Kaeng Krachan National Park, Petchaburi Province. **Regional Range:** Endemic to Thailand.

128

129 Butler's Forest Skink *Sphenomorphus butleri*

Snout to Vent: 4.3 cm. Tail unknown.

Mid-brown above, spotted with yellowish; a dark lateral band, edged with yellowish, is most distinct on the head and neck; lower part is pale yellow. Snout obtuse; lower eyelid scalely; no supranasal; rostral forming a straight suture with frontonasal; prefrontals forming a median suture; frontal shorter than frontoparietals and interparietal together; four supraoculars; parietals forming a suture behind interparietal; no nuchals; 4th and 5th labials below eye. 30-32 smooth scales round the body. **Behaviour:** Diurnal and terrestrial. Found under tree bark. **Habitat:** Presently known from peninsular monsoonal evergreen forest. **Range:** Recorded at a single site—Khao Rama, in Nakhon Si Thammarat Province in peninuslar Thailand. **Regional Range:** Thailand and Peninsular Malaysia.

129

130 Helen's Forest Skink *Sphenomorphus helenae*

Snout to Vent: 2.8 cm. Tail: 3.0 cm.

A very small forest skink. Above yellowish-brown with scattered dark brown spots. It has a dark-brown band beginning at the tip of the snout and continuing along the side of the head, which widens considerably at the shoulder; from the shoulder onward the band is broken up by short, light-brown, transverse spots. It also has a median dorsal stripe. The venter is white. The labials and sides of head, body, and tail spotted with minute grey speckles. Differs from very similar Indian Forest Skink in having three enlarged pairs of nuchals; shorter limbs which when adpressed, just overlap. There are 30 scale-rows around the mid-body. **Behaviour and Habitat:** Nothing is known of the natural history of this species. **Range:** Recorded in Nonthaburi Province in central Thailand. **Regional Range:** Endemic to Thailand.

131 Taylor's Forest Skink *Sphenomorphus lineopunctulatus*

Snout to Vent: 8.4 cm. Tail: 12.7 cm.

A medium-sized forest skink. Dark olive-brown above with irregular rows of black dots on the back. The forehead is darker than the body and covered with small black dots. It has a broad black lateral stripe punctuated at intervals with cream dots beginning on the shoulder, but preceded by one or two separate dots; this lateral stripe is bordered above by a line of ground colour and this in turn is bordered above by a very narrow, often broken, dotted brownish-white line. The chin, throat, and breast are dirty white. The venter and sub-caudal region are white. It has 38 scale-rows around the middle of the body. **Behaviour and Habitat:** Nothing is known of the natural history of this species. **Range:** Known from one record in Ubon Ratchathani Province, in northeastern Thailand. **Regional Range:** Endemic to Thailand.

132 Indian Forest Skink *Sphenomorphus indicus*

Snout to Vent: 9.7 cm. Tail: 15.4 cm.

A large forest skink. It has either a uniformly brown body, or brown with small brown or black spots arranged in longitudinal lines. It has a dark brown or black lateral stripe without flecks or spots, running from the snout along the sides of the body and extending onto the tail; above this stripe is edged with white. The head is brownish or olive-brown. The body is smooth scaled. The tail is lighter than the body, with a median and two dorso-lateral lines composed of black flecks which continue to the tip. The chin, venter, and underside of the tail are clean white. It has 34 scale-rows around the middle of the body. **Behaviour:** Diurnal and terrestrial, living in leaf litter as well as in fallen trees and branch debris. **Habitat:** Occurs in lowland evergreen forests as well as sub-montane forest at higher altitudes. **Range:** Found in forested areas throughout the country. **Regional Range:** India, Nepal, Bangladesh, Myanmar, China, Laos P.D.R., Vietnam, Cambodia, Thailand, Peninsular Malaysia, and Japan.

132

133 Larut Forest Skink *Sphenomorphus praesignis*

Snout to Vent: 10.9 cm. Tail: 13.5 cm.

A large robust forest skink, reaching more than 20 cm in total length. It has a smooth variegated buff-brown to deep-brown body with black mottling. It has 5-6 large black patches on the sides of the neck and flanks. Tail has more obvious banding of dark or blackish-brown and light brown. The throat, venter, and base of tail cream or yellow. Arms and legs also spotted. It has 3 pairs of nuchals. It has a pair of enlarged preanals. **Behaviour:** Terrestrial. **Habitat:** Peninsular monsoonal evergreen forest. **Range:** Presently known only in Khao Wang Hip, Nakhon Si Thammarat Province in peninsular Thailand. **Regional Range:** Thailand and Peninsular Malaysia.

133

134 Starry Forest Skink *Sphenomorphus stellatus*

Snout to Vent: 5.7 cm. Tail: 5.9 cm.

A small forest skink. Greenish to bronze-brown body, with characteristic white star-like markings all over; the tail is slightly lighter. It sometimes has a median black stripe and a lateral black stripe running from the eye, along the sides of the body, and continuing along the sides of the tail as a series of dark vertical bars. This black stripe also has two series of small whitish spots. The venter is greenish-white. Has well-developed limbs. It has two or three pairs of nuchals. It has 22-24 scale-rows around the middle of the body. **Behaviour:** Found in dead bark of standing trees, and in leaf litter. **Habitat:** Found in dry evergreen and southeastern monsoonal evergreen forest. **Range:** Recorded from Phu Wiang, Khon Kaen Province in northeastern Thailand and Khao Sa Bab, Chanthaburi Province in southeastern Thailand. **Regional Range:** Vietnam, Cambodia, Thailand, and Peninsular Malaysia.

134

135 Selangor Forest Skink *Sphenomorphus scotophilus*

Snout to Vent: 3.5 cm. Tail: Unknown.

A small slender forest skink. Dorsum is dark brown, with a series of small brownish-black and yellow spots which may have two prongs on the lower sides. The lips and sides of neck are spotted with brown. The venter is whitish. There are no nuchals. It has four (sometimes five) supraoculars. There are no supranasals. The lower eyelid is scaly. There are 34 scale-rows around the middle of the body. **Behaviour:** Diurnal and arboreal, being associated with tree trunks. **Habitat:** Presently known only from peninsular monsoonal evergreen forest, though may be expected to occur in Malayan mixed dipterocarp forest. **Range:** Currently only recorded in the Khao Luang range, Nakhon Si Thammarat Province in peninsular Thailand. **Regional Range:** Thailand and Peninsular Malaysia.

135

Supple Skinks Genus *Lygosoma*

These skinks have elongated bodies and pentadactyl limbs which are greatly reduced in size and widely separated. The tympanum is often small and deeply sunken. Eyelids are reduced in some species. Eight species are recorded in Thailand.

KEY TO THE THAI *LYGOSOMA*

1	**a**	Scales smooth	**2**
	b	Dorsal scales finely keeled	***L. herberti***
2	**a**	A dark flank stripe present	**3**
	b	No dark flank stripe	**4**
3	**a**	Frontoparietal is single	***L. frontoparietale***
	b	Two frontoparietals	***L. bowringii***
4	**a**	Lower eyelid with a disc	***L. anguinum***
	b	Lower eyelid scaly	**5**
5	**a**	Forelimb is 3.5-4.5 times the length of axilla to groin; second toe is much shorter than third	**6**
	b	Forelimb is 7-12 times the length of axilla to groin; second toe not or barely shorter than third and fourth	**7**
6	**a**	Supranasal fused to nasal in front of nostril; ear opening with 2-3 lobules	***L. koratense***
	b	Supranasal free of nasal; ear opening without lobule	***L. corpulentum***
7	**a**	Head dark, with pale blotches on side of neck or bars across the nape	**8**
	b	No contrasting neck markings, dorsal surface plain or with dark stripes	***L. quadrupes***
8	**a**	Supranasals separated on the midline not fused to nasal	***L. isodactylum***
	b	Supranasals in contact on the midline, not fused to nasal	***L. haroldyoungi***

Key modified from Dring (unpublished).

136 Bowring's Slender Skink *Lygosoma bowringii*

Snout to Vent: 4.7 cm. Tail: 5.5 cm.

Bronze brown body edged by a diffuse pair of light brown or cream dorso-lateral stripes. These stripes are bordered by a distinctive, narrow black stripe running to the groin. Chin and venter are ivory yellow. The tail is about 1.25 times the body length. **Behaviour:** Diurnal. Terrestrial and sub-terrestrial, being found in the leaf litter and loose soil. Diet consists of small insects. Oviparous, laying 2-4 eggs. **Habitat:** Recorded from all forest types below 1,500 m, as well as other habitats including stream beds and gardens. **Range**: Recorded from all regions. Very abundant. **Regional Range:** India (Andamans and Nicobars), Myanmar, Vietnam, Cambodia, Thailand, Peninsular and Eastern Malaysia, Singapore, and Indonesia (Sumatra and Java).

136

137 Herbert's Slender Skink *Lygosoma herberti*

Snout to Vent: 6.6 cm. Tail: 7.1 cm.

Larger-slender bodied but proportionately stouter skink than Bowring's Slender Skink. Dorsal coloration is light brown, similar to Bowring's Slender Skink. Dark dorso-lateral stripe varies in distinctness and the distance to which it extends along the side of the body. Venter is light brown. **Behaviour:** Found in foothills, but natural history is largely unknown. **Habitat:** Found in peninsular monsoonal evergreen forest. **Range:** Confined to peninsular Thailand. Recorded in Ranong, Phang-nga, Phuket, Surat Thani and Nakhon Si Thammarat provinces. **Regional Range:** Endemic to Thailand.

137

138 Even-toed Supple Skink *Lygosoma isodactylum*

Snout to Vent: 8.2 cm. Tail: Up to 6.6 cm.

Back dark olive. Chin, venter, and underside of tail yellowish. The ratio of the axilla-to-groin length to the arm length is approximately 7.25 times. There are 30-34 scale-rows around the middle of the body, and 88-98 scales in a line between the parietals and a point above vent. **Behaviour:** Sub-fossorial, living in cracks in the ground. Fast snake-like movements. **Habitat:** Found in open fields and gardens in the lowlands. **Range:** Recorded in Nakhon Sawan, Saraburi, and Nakhon Si Ayuthaya provinces in central Thailand. **Regional Range:** Laos, Cambodia, and Thailand.

138

139 Khorat Supple Skink *Lygosoma koratense*

Snout to Vent: 10.6 cm. Tail: 9.5 cm.

Violet-brown elongate supple skink with lighter sides and a greyish-white venter. Distance between the end of the snout and arm insertion, twice the axilla-to-groin distance or slightly more. Short limbs when adpressed are widely separated. The thick tail is shorter than the head and body. There are 32-34 scale-rows about middle of the body. 60-63 scales from parietals to above vent. **Behaviour:** Burrowing nature, living in debris. **Habitat:** Open fields. **Range:** Restricted to Saraburi and Nakhon Ratchasima provinces in central Thailand. **Regional Range:** Endemic to Thailand. **International Conservation Status:** IUCN: Endangered.

139

140 Annamese Supple Skink *Lygosoma corpulentum*

Snout to Vent: 16.5 cm. Tail: 15.0 cm.

Light yellowish-brown or chocolate brown, thickly mottled on the back and sides with dark brown; lips, sides of neck, and throat yellowish; labial shields edged with black. Distance between the end of the snout and the forelimb 2-2.20 times the distance between the axilla and groin; snout obtuse; lower eyelid scaly; supranasals entire, in contact with one another; frontal longer than the fronto-parietals and interparietal together; no nuchals; no enlarged temporals. Body scales smooth, dorsals a little larger than the laterals or ventrals, 36 to 38 scales at the middle. **Behaviour:** Inhabits wet evergreen forest. **Habitat:** Occurs in southeastern monsoonal evergreen forest. **Range:** Recorded in Khao Ang Rue Nai Wildlife Sanctuary, Chachaengsao Province and Khao Sao Doi Wildlife Sanctuary, Chanthaburi Province in southeastern Thailand. **Regional Range:** Vietnam, Cambodia, and Thailand.

140

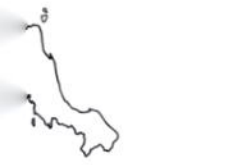

141 Burmese Supple Skink *Lygosoma anguinum*

Snout to Vent: 5.5 cm. Tail: 6.2 cm.

A Supple Skink with short limbs, a slender body, and a tail about as long as the head and body. It is brown above, paler below; a dark-brown dorsolateral streak more or less distinct. There are 22 scales around the body. **Behaviour:** Found in loose soil and under piles of rotting grass. Nothing else is known of its natural history. **Habitat:** Found in dry savannah. **Range:** Presently appears to have a disjunct distribution, being found in northern Thailand and upper peninsular Thailand. Recorded in Mae Hong Son, Prachuap Khiri Khan, and Chumphon provinces. **Regional Range:** Myanmar and Thailand.

141

142 Banded Supple Skink *Lygosoma haroldyoungi*

Snout to Vent: 13.6 cm. Tail: At least 7.5 cm.

A supple skink with a greatly elongated, dull-black body; an ivory-white ring is present about the head behind the parietals, while the body has 27 irregular ivory-coloured bands. The tail has 33 bluish bands. The venter is pale yellow with irregular spots and markings. **Behaviour:** Sub-fossorial, feeding on earthworms and other insects. Sometimes found in termite mounds or under rotten logs. **Habitat:** Found in sub-montane forests. **Range:** Northern and upper northeastern Thailand, recorded from Chiang Mai, Loei, Chaiyaphum, and Sakhon Nakhon provinces. **Regional Range:** Laos and Thailand. **International Conservation Status:** IUCN: Endangered.

142

143 Common Supple Skink *Lygosoma quadrupes*

Snout to Vent: 6.8 cm. Tail: 6.9 cm.

A very elongate, diminutive skink with very reduced limbs, but which still has five digits. The colour is grey brown to brown, with dark longitudinal lines usually visible on the edges of all scale-rows, which continue onto the tail. Forehead and supralabials are darker than the rest of the dorsum. Venter and sub-caudals are pale pink. There are 24-26 scale-rows around the middle of the body. **Behaviour:** Sub-fossorial, being found in leaf litter. Feeds on termites. Two to three eggs per clutch. **Habitat:** Found in a wide range of forest types. **Range:** Found throughout Thailand, but localized. **Regional Range:** China, Vietnam, Laos, Cambodia, Thailand, Peninsular and Eastern Malaysia, Singapore, Indonesia (Sumatra and Java), and the Philippines.

143

Larut Skinks Genus *Leptoseps*

This genus comprises members from Malaysia and Sumatra, including Larut Province in Peninsular Malaysia. The body is very elongate while the limbs are very small. A single species is present in Thailand.

144 **Boehme's Larut Skink** *Leptoseps osellai*

Snout to Vent: 4.1 cm. Tail: 5.2 cm.

Elongate serpentine, body with head slightly broader than the neck. Upper side of body and tail is brownish with a pattern of darker brown longitudinal stripes; the median stripes are broader than the outer stripes. The strongly reduced limbs have four digits or toes, a characteristic feature not currently shared by any other Southeast Asian skink. The eyes are relatively large. The snout is slightly pointed. The venter is cream. Mid-body has 18 scale-rows. **Behaviour:** Terrestrial, found under logs and other debris. **Habitat:** Sub-montane and montane forests, including secondary forest. **Range:** Known from two sites in northern and northeastern Thailand; Mae Koung, Chiang Mai Province and Nam Nao National Park, Petchabun Province. **Regional Range:** Endemic to Thailand.

144

Legless Skinks Genus *Isopachys*

Members of this genus are vermiform, lacking limbs and having only vestiges of pectoral and pelvic girdles. The head is continuous with the body, lacking any external neck segment. The snout region is covered by a series of enlarged heavily keratinized shields to facilitate burrowing. The eyes are vestigial. The tympanum is encapsulated and not visible externally. Four species occur in Thailand.

KEY TO THE THAI *ISOPACHYS*		
1 a	Nasal scales separated; frontoparietals in contact	*I. anguinoides*
b	Nasal scales in contact; frontoparietals separated	**2**

2 a	Longitudinal markings; dark venter; 20-22 scale-rows around mid-body; 5 supralabials; snout to vent 77-177 mm	*I. borealis*
b	4 supralabials	**3**
3 a	18 scale-rows around mid-body; snout to vent 82-106 mm	*I. roulei*
b	Longitudinal markings; 24-28 scale-rows around mid-body; snout to vent 167-220 mm	*I. gyldenstolpei*

Key modified from Taylor (1963); Lang & Boehme (1990).

145 Striped Legless Skink *Isopachys anguinoides*

Snout to Vent: 7.1 cm. Tail: 6.2 cm.

Body is very elongate and limbless. Pale greyish-brown with a pair of dark brown streaks along the middle of the back, and a broader dorso-lateral pair. Below these, longitudinal lines and the ventral surface are darker. Nasal shields are separated from or just touching one another; in Roule's Legless Skink the contact is much broader. There are 22-24 scales round the body. **Behaviour:** Sub-fossorial, usually found under rotting debris or piles of rotting grass. Often burrow rapidly into soft earth. **Habitat:** Found near coast in dry sandy environments. **Range:** Found in the upper part of peninsular Thailand, being endemic to the country. Recorded in Petchaburi, Prachuap, Khiri Khan, and Chumphon provinces. **Regional Range:** Endemic to Thailand.

145

146 Roule's Legless Skink *Isopachys roulei*

Snout to Vent: 9.5 cm. Tail: 11.4 cm.

Very elongate and limbless. Head and back are slightly pinkish-fawn with two broad dark bluish-brown longitudinal stripes running along the length of the body. Ventral surface is brown. Top of head is mostly dark brown. The nasal shields are in broad contact with one another; in the Striped Legless

Skink they are barely touching. **Behaviour:** Sub-fossorial. Usually found under rotting debris or piles of rotting grass. Often burrow rapidly into soft earth. **Habitat:** A single specimen was found in sandy soil in a coconut grove. **Range:** Known from a single site at Ang Sila, Bang Saen, Chonburi Province in southeastern Thailand. **Regional Range:** Endemic to Thailand.

146

147 Western Legless Skink *Isopachys borealis*

Snout to Vent: 15.4 cm. Tail: 3.7 cm.

Very elongate body, and limbless. Distinguished from all other legless lizards by its dark brown venter and its unique dorsal pattern; the dorsal markings have an irregular outline and are round. Has 44 pairs of dark brown spots on an off-white background. By comparison, similar Gyldenstolpe's Legless Skink has dorsal markings which are extremely regular and rectangular. Both Striped and Roule's Legless skinks have longitudinal markings. Around the neck are 24 longitudinal scale-rows, which reduce to 22 around mid-body and 15 at base of tail. Tip of snout slightly more yellow than dorsal coloration. **Behaviour:** Fossorial, burrowing through topsoil. **Habitat:** Moist topsoil of cultivated lands (sweet potato and pineapple) in clearings with moderate sunlight. Occurs from sea level up to 300 m. **Range:** Northeastern and western Thailand. Recorded in Petchabun, Khon Kaen, Kamphaeng Phet, Tak, and Uthai Thani provinces. **Regional Range:** Myanmar and Thailand.

147

148 Gyldenstolpe's Legless Skink *Isopachys gyldenstolpei*

Snout to Vent: 22.0 cm. Tail: 11.4 cm.

Very elongate body. The terminal portion of the body and dorsal part of the tail have a checkered pattern as in Western Legless Lizard but the markings are extremely regular and rectangular. The rest of the dorsum is dark brown with yellow longitudinal stripes, 3-4 scales wide. **Behaviour:** Usually found under rotting debris or piles of rotting grass. Often burrows rapidly into soft earth. Diet probably consist of earthworms. **Habitat:** Dry sandy environments. **Range:** Recorded from Koh Lak, Hua Hin, Prachuap Khiri Khan Province, and Kanchanaburi Province. **Regional Range:** Endemic to Thailand. **International Conservation Status:** IUCN: Data Deficient.

148

Legless Skinks Genus *Davewakeum*

This genus comprises a single skink species occurring in Thailand.

149 Miriam's Legless Skink *Davewakeum miriamae*

Snout to Vent: 11.4 cm. Tail: 17.3 cm.

Slender-bodied, legless skink in which head is not differentiated from rest of the body. The head and body are brown coloured with a darker series of brown lines caused by spots on each scale; the sides and venter may be slightly lighter. Scales are smooth. Twenty-two scale-rows around mid-body. The tip of the tail is blunt. **Behaviour:** Nocturnal. Usually in forest debris, under rocks or in the loose soil down to a depth of 15 cm. May be found on the ground's surface. **Habitat:** Inhabits dry evergreen forest. **Range:** Central Thailand. Recorded in Wang Nam Khieo, Sakaerat, Nakhon Ratchasima Province. **Regional Range:** Endemic to Thailand.

149

Snake Skinks Family Dibamidae

Members of this family have vermiform bodies. The eyes are concealed under the skin. The forelimbs are absent. The hind limbs are represented only in the male by a pair of short, scaled flipper-like appendages, lying in a deep groove on either side of the vent. Only one genus, *Dibamus*, represented by 2 species, is known from Thailand.

Snake Skinks Genus *Dibamus*

The snout is covered with a thickened shield which is more or less divided by suctures. Four shields on the top of the head. There is no ear opening. The tail is short and obtuse.

KEY TO THE THAI *DIBAMUS*

1 a Snout sub-conical covered by a large rostral, with an entrant suture from ocular at about level of eye; body scales at mid-body 20 — *D. alfredi*

b Snout bluntly round, rostral suture complete and joining to form a single median suture just above the tip of snout; body scale at mid-body 19 — *D. somsaki*

Key prepared by Chan-ard, Parr and Nabithabhata.

150 **Alfred's Snake Skink** *Dibamus alfredi*

Snout to Vent: 9.9 cm. Tail: 1.7 cm.

Body is generally violet to purplish-brown, being slightly lighter on the ventral surface. Has creamy-white to ivory-white underside of snout, area around nostrils, sides of head, and anterior part of chin. The dorsal surface of limbs is white. The snout is sub-conical and covered by a large rostral. There are 20-21 scale-rows around the body. **Behaviour:** Found in humus or leaf litter. **Habitat:** Found in peninsular monsoonal evergreen forest and Malayan mixed dipterocarp forest. **Range:** Recorded in Bukit Besar, Pattani Province and Yala Province in extreme southern peninsular Thailand. **Regional Range:** Thailand, Peninsular Malaysia, and Indonesia (Nias Island).

150

151 Khao Soi Dao Snake Skink ***Dibamus somsaki***

Snout to Vent: 10.6 cm. Tail: 2.5 cm.

Extremely elongate and legless. Dorsum, venter, tail, and male hindlimbs are uniformly purplish brown; the tip of the snout and lower jaw are slightly lighter. Differs from Alfred's Snake Skink in having rostral sutures complete but joining to form a single median suture just above the tip of the snout. Snout is bluntly rounded. Mid-body scale-rows number 18-19. Sub-caudals number 57-59. **Behaviour:** Fossorial. Found in humus or leaf

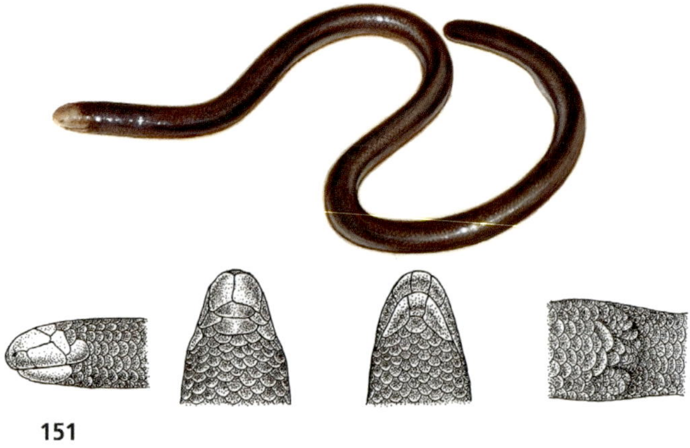

151

litter. **Habitat:** Found in secondary forest in southeastern monsoonal evergreen forest. **Range:** Southeastern Thailand. Recorded in Khao Soi Dao Wildlife Sanctuary, Chanthaburi Province. **Regional Range:** Endemic to Thailand.

Glass-Lizards Family Anguidae

Members of this family have their scales reinforced with bony plates called osteoderms, a characteristic shared with the skinks. However, the presence of a deep flexible groove that runs along each side of the body allows expansion of the body when breathing or when the body is distended with food or eggs. The legless Glass Lizards can easily be mistaken for snakes. Unlike snakes, they possess movable eyelids and also have a distinct ear opening. Their tails are very long and are very fragile. They are good burrowers.

Glass Lizards Genus *Ophisaurus*

A single member of this genus is found in Thailand.

---

152 Asian Glass Lizard *Ophisaurus gracilis*
Snout to Vent: 35.0 cm. Tail: 54.0 cm.

Very long, rich reddish-brown lizard with pale blue transverse bars on upper part of the dorsum. Venter is orange. The pupil is black. **Behaviour:** Found in the leaf litter. Sometimes found basking in the early morning. **Habitat:** Diurnal and terrestrial. Found in stands of rhododendrons in sub-montane and montane forests on isolated mountain peaks above 1400 m. **Range:** Confined to northern and northeastern Thailand. **Regional Range:** India, Myanmar, China, Vietnam, and Thailand.

152

Monitors Family Varanidae

Members of this family are very large lizards, usually 1-3 m in length. The head has numerous small scales. The tongue is long, slender, and deeply bifid and can be retracted into a sheath. The body is covered in small tuberculate juxtaposed scales, each surrounded by a row of minute tubercles. They are always pentadactyl. They have an elongate tail. Monitors will eat almost any animal that they can capture, and appear to be very fond of carrion. Four species are found in Thailand.

KEY TO THE THAI VARANUS

1 a	Enlarged scales on dorsal surface of neck rather widely spaced; twice as far from end of snout as from orbit	**2**
b	Scales on neck not widely spaced nor especially enlarged; but nearer to end of snout than to orbit	**3**
2 a	Enlarged flat neck scales arranged in 10-12 longitudinal rows, the posterior ones keeled; dorsal scales strongly keeled; blackish above, neck and anterior part of body yellowish with three black longitudinal stripes on head and neck; two black bands across shoulders	*V. rudicollis*
b	Enlarged flat neck scales not arranged in distinct rows, dorsal scales slightly enlarged, strongly keeled; snout depressed at end; head light brown to yellowish-brown with a pair of black streaks from eye joining a transverse dark band on neck; back with four rather distinct broad darker bands narrowly separated by light dotted bands, with one on neck wider and more distinct	*V. dumerilii*
3 a	Snout depressed at end, half as long as head; nostril rounded or oval, twice as far from orbit as from tip of snout; usually have five or six transverse bands of yellow ocellated spots across back; head banded anteriorly with black; ventrals keeled.	*V. salvator*
b	Snout not depressed at end; nostril not twice as far from orbit as from end of snout; ventral scales smooth; olive black on body with numerous small yellow spots not forming well-defined crossbars; a dark temporal stripe from eye merging with dark bands on neck	*V. bengalensis*

Key modified from Taylor (1963).

153 Red-necked Monitor *Varanus rudicollis*

Snout to Vent: 30.6 cm. Tail: 42.6 cm.

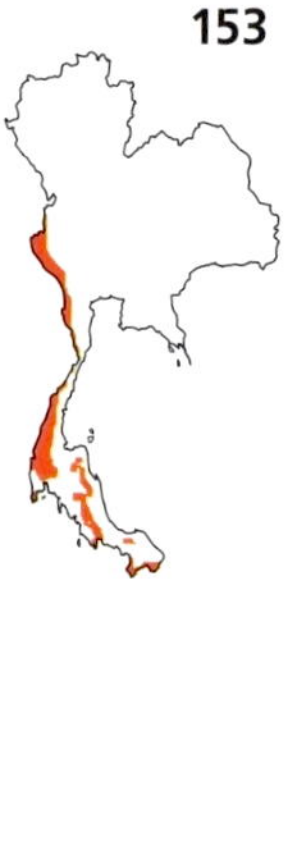

Blackish above, neck and anterior part of body yellowish with three black longitudinal stripes on head and neck. Two black bands across shoulders. Body has yellow, transverse rows of ocelli. Enlarged flat neck scales arranged in 10-12 longitudinal rows; in Harlequin Monitor they are not arranged in rows. The posterior neck scales are keeled. Sides of neck reddish. **Behaviour:** Arboreal. **Habitat:** Found in peninsular monsoonal evergreen forest and Malayan mixed dipterocarp forest. **Range:** Found throughout peninsular Thailand northward up the Tenasserim range until Tak Province. **Regional Range:** Myanmar, Thailand, Peninsular and Eastern Malaysia, and Indonesia (Sumatra).

153

154 Harlequin Monitor *Varanus dumerilii*

Snout to Vent: 98 cm. Tail: 114 cm.

Head is light brown to yellowish-brown with a pair of black streaks from the eye joining a transverse dark band on the neck. The back has four rather distinct broad darker bands narrowly separated by light dotted bands. Enlarged flat neck is not elongated. Snout depressed at the end. Nostrils twice as close to the orbit as the end of the snout. The nuchal scales are larger than those found on the crown on the head. **Behaviour:** Arboreal. Perhaps riparian. **Habitat:** Found in peninsular monsoonal evergreen forest and Malayan mixed dipterocarp forest. **Range:** Found throughout peninsular Thailand northward up to Sai Yok National Park in Kanchanaburi Province. **Regional Range:** Myanmar, Thailand, Peninsular and Eastern Malaysia, and Indonesia (Kalimantan and Sumatra). **International Conservation Status:** CITES: Appendix II.

154

155 Water Monitor *Varanus salvator*

Snout to Vent: 98 cm. Tail: 114 cm.

A very large monitor, reaching in excess of 2.5 m. Body has five or six rows of round, yellow, often ocellate spots. Belly is yellow. The tail is banded with black and yellow along its length. The snout is long, three times as long as high, with the nostrils round or oval, situated twice as far from the eye as from the tip of the snout. **Behaviour:** More aquatic than other monitors, being seldom found far from water. **Habitat:** Frequents rivers, canals, and mangroves. **Range:** Found throughout Thailand, although quite rare in the northeast. **Regional Range:** India, Sri Lanka, Myanmar, China, Vietnam, Laos, Cambodia, Thailand, Peninsular and Eastern Malaysia, Singapore, and Indonesia (Kalimantan, Sumatra, Java, and Sulawesi).

155

156 South-east Asian Monitor *Varanus nebulosus*

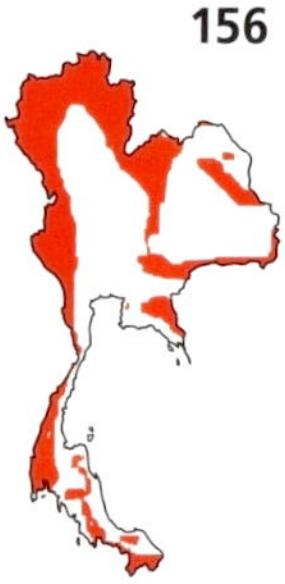

Snout to Vent: 220 cm. Tail: 348 cm.

Dorsal surface is olive-black with numerous yellow scales scattered over the dorsum, sides, arms, and legs. On the back, the larger yellow spots are usually surrounded by back scales forming ocelli in indefinite rows. The chin has transverse bars of yellowish-white. The tail has two broad bands of black separated by narrow yellow bands. The nostrils are located near to snout. Young animals are olive to black and the chin is heavily barred with blue and yellowish-cream. **Behaviour:** Usually found on forest floor but climbs trees to escape, often hiding in tree holes. **Habitat:** Dry evergreen forest, mixed deciduous forest, and dry dipterocarp forest. **Range:** Found in forested areas throughout the country; rare in the northeast. **Regional Range:** Myanmar, Vietnam, Thailand, Peninsular Malaysia, and Indonesia (Java).

156

Blind Snakes Family Typhlopidae

The typhlopids are small snakes, usually smaller than 30 cm long. Due to their lifestyle of burrowing in loose soil, leaf litter, and rotting logs, their eyes are vestigial. They are nocturnal. They have blunt, indistinct heads, cylindrical bodies of uniform diameter and short blunt tails. The body scales are smooth, shiny, and uniform. Blind Snakes lack enlarged ventral scales as found in other snakes. The pelvis is usually present as a single bone or cartilage, but there is no evidence of external limbs. The family is characterized by the maxillary bone—bearing teeth and arranged transversely—which can be rotated slightly on its attachment to the skull; the dentary bone usually lacks teeth (present in anomalepine species). Circumtropical in distribution. Worldwide the family comprises 6 genera, of which 12 species from the genera *Ramphotyphlops* and *Typhlops* are found in Thailand.

SUBORDER SERPENTES

Snakes are elongated, legless, carnivorous reptiles of the suborder Serpentes that can be distinguished from legless lizards by their lack of eyelids and external ears. Like all squamates, snakes are ectothermic, amniote vertebrates covered in overlapping scales. Many species of snakes have skulls with several joints, enabling them to swallow prey much larger than their heads with their highly mobile jaws. To accommodate their narrow bodies, snakes' paired organs appear one in front of the other instead of side by side, and most have only one functional lung. Some species retain a pelvic girdle with a pair of vestigial claws on either side of the cloaca.

KEY TO THE THAI *RAMPHOTYPHLOPS* AND *TYPHLOPS*

1 a	Preocular fused to nasal, inferior nasal suture contacting first supralabial, rostral broad (greater than 0.6 head width), dorsum striped	*R. lineatus*
b	Preocular separate from nasal, inferior nasal suture contacting second supralabial or preocular, rostral narrow (less than 0.4 head width), dorsum uniform	**2**
2 a	Mid-body scale-rows number 18	**3**
b	Mid-body scale-rows number 20-30	**4**
3 a	Glands present under central portions of the head shield, sub-ocular present, 2-3 postoculars, sub-caudals more than 15	*T. floweri*
b	Glands confined to scale sutures, not present under central portions of shields, lacking sub-ocular, 1 postocular, sub-caudals fewer than 15	*T. porrectus*
4 a	Mid-body scale-rows number 20-22	**5**
b	Mid-body scale-rows number 24-30	**10**
5 a	Inferior nasal suture contacting preocular	**6**
b	Inferior nasal suture contacting second supralabial	**7**
6 a	Eye with visible pupil, under ocular, snout not depressed, prefrontal, frontal, and interparietal enlarged	*R. braminus*
b	Eyespot, lacking pupil, under supraocular-ocular border, snout depressed, prefrontal, frontal and parietal enlarged	*T. khoratensis*
7 a	Superior nasal suture extending dorsally, parallel to rostral and visible from above	**8**
b	Superior nasal suture extends horizontally to rostral, not visible dorsally	**9**
8 a	Head, neck, and anal scales whitish, tail tip whitish. Glands beneath the scales rarely seen. Superior nasal suture contacts the rostral	*T. albiceps*
b	Only snout whitish; from anal scale to tail tip whitish. Glands beneath the scales clearly seen. Superior nasal suture not contact the rostral	*R. osakiae*
9 a	Mid-body scale-rows 20, postnasal deeply concave	*T. roxaneae*
b	Mid-body scale-rows 20-22, postnasal not concave	*T. siamensis*
10a	Mid-body scale-rows 24, lacking scale-row reduction, eye indistinct, parietals barely broader than deep, dorsal head scales large and uniform	*T. trangensis*
b	Mid-body scale-rows 24-30, mid-body and/or posterior scale-row reduction of at least two rows, eye distinct with pupil, parietals at least twice as broad as deep, dorsal head scales variable in size	**11**
11a	Sharp demarcation between dark-brown, purple or black dorsum and yellow or gold venter	*T. muelleri*
b	Gradual lateral transition between brown dorsum and light-brown venter	*T. diardi*

Key modified from Niyomwan (1999); Wallach (2001).

Blind Snakes Genus *Ramphotyphlops*

Members of this genus are small burrowing snakes. Four species are found in Thailand.

157 White-headed Blind Snake *Ramphotyphlops albiceps*

Body Length: Up to 18.3 cm.

This blind snake is very small. It has diagnostic coloration, with a brown body and white head, anterior part of neck, tail, and anal region. The 6-8 scale-rows on the ventral surface are creamy white. There are 20 longitudinal scale-rows around the body from head to tail. The total length is about 39 times the body width. The ocular touches the 3rd and 4th labials. **Behaviour:** Sub-fossorial. **Habitat:** Found in a wide range of primary forest habitats in the lowlands. **Range:** Found throughout Thailand apart from the northeast. Recorded in Chiang Mai, Nakhon Ratchasima, Ratchaburi, Chanthaburi, Nakhon Si Thammarat, Pattani, and Yala provinces. **Regional Range:** Myanmar, China, Thailand, and Peninsular Malaysia.

157

158 Common Blind Snake *Ramphotyphlops braminus*

Body Length: Up to 18.0 cm.

A small blind snake with an eye with a visible pupil under the ocular. Dark brown or black above and paler brown below. The snout, which is not depressed, is pale as is the tip of the tail. It has 20 scale-rows around the mid-body. This is the only known parthenogenic snake species; all individuals are females—males are unknown. **Behaviour:** Nocturnal and sub-fossorial, being found in leaf litter, as well as in rotten logs. Active on the surface, especially after rains. Diet includes ants and termites, and their larvae. **Habitat:** Found in a wide range of forested habitats, from lowlands up to the top of mountains. Found near human settlements. **Range:** The species probably occurs throughout Thailand. Records from Chiang Mai, Phrae, Udon Thani, Prachinburi, Chanthaburi, Chonburi, Saraburi, Pathum Thani, Prachuap Khiri Khan, Phatthalung, and Songkhla provinces. **Regional Range:** Sub-Saharan Africa, Middle East, Indian Ocean Islands, South Asia and Southeast Asia, Australia, Pacific Islands, Mexico (introduced), Central America (introduced), and North America (introduced).

158

159 **Striped Blind Snake** *Ramphotyphlops lineatus*

Body Length: Up to 48.0 cm.

The largest blind snake. It has a diagnostic body coloration, being pale brownish above, with a yellow spot on each scale forming longitudinal lines along the body. The head and venter are yellowish-white. The snout is rounded and projects strongly. The eyes are hidden. The preocular is fused

159

to the nasal. The ocular touches the 2nd and 3rd labials. The rostral is large, being more than 60% of the width of the head. It has 22 scale-rows around the mid-body. **Behaviour:** Like all blind snakes, has a burrowing habit. Feeds on ant or termite pupae, larvae, eggs, and occasionally adults. **Habitat:** Inhabits Malayan mixed dipterocarp forest. **Range:** Southernmost provinces of peninsular Thailand. Recorded in Na Pradoo, Pattani Province and Muang District, Narathiwat Province. Considered uncommon. **Regional Range:** China, Vietnam, Laos, Cambodia, Thailand, Peninsular Malaysia, Singapore, and Indonesia (Kalimantan, Sumatra, and Java).

160 Ozaki's Blind Snake *Ramphotyphlops ozakiae*

Body Length: Up to 15.1 cm.

A small blind snake, with an eye visible under the ocular. Dark brown or black above and paler below. The snout is white and the anal plate to the tail tip are whitish, which are characteristic features. The glands can be clearly seen beneath the scales. It has 20 scale-rows around the mid-body. **Behaviour:** Burrowing habit, being found in leaf litter, as well as in rotten logs. **Habitat:** Found in dry evergreen forest. **Range:** Only occurs in Nakhon Ratchasima Province, northeastern Thailand. **Regional Range:** Endemic to Thailand.

Blind Snakes Genus *Typhlops*

Members of this genus are small burrowing snakes. Eight species are found in Thailand.

161 Flower's Blind Snake *Typhlops floweri*

Body Length: Up to 22.8 cm.

One of the smaller blind snakes, with 18 longitudinal scale-rows around the body. The black or very dark brown body is slender and elongate, the length being 85-90 times the diameter. The sub-caudals number more than 15. The head is slightly flattened. It has a projecting, rounded snout, the tip of which is yellowish. Glands are present under the central portions of the head shields. The nasal is completely divided. A sub-ocular is wedged between the lower part of the ocular and preocular. The venter is pale brown. **Behaviour:** Presumed to have a burrowing habit. **Habitat:** Habitat requirements not clearly understood. **Range:** Confined to central Thailand. Recorded in Bangkok and Pathum Thani Province. **Regional Range:** Endemic to Thailand.

161

162 Khorat Blind Snake *Typhlops khoratensis*

Body Length: Up to 13.0 cm.

The smallest of the Thai blind snakes. Uniformly grey apart from the scales around the mouth and underside of snout which are whitish. It has an eye-spot which lacks a pupil. The snout is depressed. It has two pairs of parietals; in comparison the Common Blind Snake has only one pair. There are 20 scale-rows throughout the body and 315-326 transverse scale-rows. The body length is 28-43 times the body width. **Behaviour:** Burrowing habit, found in soil, under debris, or in rotting logs. **Habitat:** Found in dry evergreen forest. **Range:** Found on the Khorat Plateau, northeastern Thailand. **Regional Range:** Endemic to Thailand.

162

163 Slender Blind Snake *Typhlops porrectus*

Body Length: Up to 28.5 cm.

A medium-sized blind snake, with 18 scale-rows around the mid-body like Flower's Blind Snake. The body is brownish-black or mid-brown above, and paler below. The total length is 50-60 times the diameter of the body; in Flower's Blind Snake the total length is about 85 times the diameter of the body. There are fewer than 15 sub-caudals. The snout, chin, and anal region are usually whitish. The snout is rounded and projects strongly. It lacks a sub-ocular and has only one postocular. **Behaviour:** Sub-fossorial, being found in soil, under debris, and in rotting logs. **Habitat:** Found in montane evergreen forest. **Range:** Recorded in Doi Pui, Chiang Mai Province. Introduced in Bangkok and Pathum Thani Province. **Regional Range:** Pakistan, India, Sri Lanka, Bangladesh, Myanmar, Thailand, and Peninsular Malaysia.

163

164 Trang Blind Snake *Typhlops trangensis*

Body Length: Up to 15.5 cm.

This species is a short, relatively thick-bodied blind snake, with 24 scale-rows at mid-body. The body length is approximately 30 times the body width. It is ultramarine blue or grey on the 11 dorsal scale-rows while the lower 13 scale-rows are creamy white, with the dividing line rather sharply defined. The dorsal head scales are large and uniform. The eye is indistinct. **Behaviour:** The only recorded individual was found under a decaying log. **Habitat:** Found in peninsular monsoonal evergreen forest. **Range:** The species is only known from a single animal found in Khao Chong, Trang Province. **Regional Range:** Endemic to Thailand.

164

165 Diard's Blind Snake *Typhlops diardii*

Body Length: Up to 40.0 cm.

One of the largest blind snakes. The nine dorsal scale-rows of the neck and the anterior part of the body are dark brown; this coloration gradually expands along the body to 11 dark brown scale-rows before the middle of the body. This dorsal coloration grades into the ventral coloration of brownish-cream. The snout is rounded and projecting. The eye is distinctly visible, with a pupil. It has a preocular. The tail terminates in a pointed spine pointing downward. It has 24-30 mid-body scale-rows. **Behaviour:** Sub-fossorial, found in soil, surface debris, and rotten logs. **Habitat:** Found in dry evergreen forest and sub-montane forests. **Range:** Found in forested areas in continental Thailand. **Regional Range:** India, Bangladesh, Myanmar, China, Vietnam, Laos, Cambodia, and Thailand.

165

166 Mueller's Blind Snake *Typhlops muelleri*

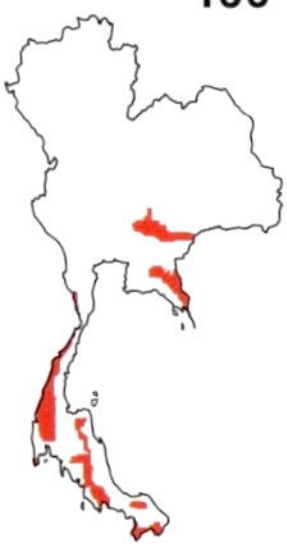

Body Length: Up to 42.0 cm.

One of the larger typhloids. It is uniformly dark brown, purple or blackish above, although each scale is darker near its centre, which contrasts with the sharp demarcation of the gold or yellow venter. In the similar Diard's Blind Snake the lateral colour transition is gradual. The head is slightly flattened. The 15 scale-rows on the lower sides and ventral surface are creamy white. The front part of the head is creamy white. The terminal point of tail is white. It has 28 longitudinal scale-rows. **Behaviour:** Burrowing habit in soft soil and surface debris. Feeds on ant and termite larvae. **Habitat:** Found in evergreen forests. **Range:** Recorded in the Dong Phaya Yen Range southward to peninsular Thailand. **Regional Range:** Myanmar, Vietnam, Laos, Cambodia, Thailand, Peninsular Malaysia, Indonesia (Kalimantan and Sumatra).

166

167 Siamese Blind Snake *Typhlops siamensis*

Body Length: Up to 16.6 cm.

A very small typhloid. Snout roundish and moderately projecting. Eyes are visible. Cylindrical body is greyish olive or brownish above and yellow below. Tail ends in a point. There are 22 scale-rows around the body. Total body length is 33 times as long as the body diameter. **Behaviour:** Presumed to be fossorial, associated with leaf litter. **Habitat:** Found in southeastern monsoonal evergreen forest. **Range:** Known from a single specimen in Chanthaburi Province, southeastern Thailand. **Regional Range:** Cambodia and Thailand.

167

168 Roxane's Blind Snake ***Typhlops roxaneae***

Body Length: Up to 23.1 cm.

A small blind snake. Golden brown dorsal coloration separated from yellowish-tan venter by a single lateral row on each side. The ventral snout is yellow, as is the cloacal region. There is a yellow ring around the tail tip. It has a small number (20) of mid-body scale-rows. **Behaviour:** Behaviour unknown, but assumed to have a burrowing habit. Inhabits soil and surface debris. **Habitat:** Found in urban areas. **Range:** Known from the lower Central Plains, with records from the vicinity of an international school at Sri Ayuthaya Road, Bangkok. **Regional Range:** Endemic to Thailand.

Pipe Snakes Family Cylindrophiidae

The family is characterized by having a generally cylindrical body of nearly equal diameter throughout, with uniform scales. The ventral scales are either not enlarged or poorly so. Vestiges of a pelvis and legs are present, the latter terminating in a claw-like appendage lateral to the vent. A single species occurs in Thailand.

Pipe Snake Genus *Cylindrophis*

One species occurs in Thailand.

169 Red-tailed Pipe Snake ***Cylindrophis ruffus***

Body Length: Up to 86.5 cm.

A medium-sized snake. It has a strongly iridescent cylindrical body, of nearly equal diameter throughout. The dorsal surface is deep lavender-purple, with whitish skin between scales. The venter has approximately 48 greyish-white bars extending onto the sides. The underside of the tail is red with considerable darker pigment. The tip is black. The body scales are small and imbricating.

169

The scales are in 21 rows. It has 9-12 teeth on the maxilla; premaxilla without teeth. **Behaviour:** Nocturnal and sub-fossorial. **Habitat:** Found in wetlands, flooded paddy fields, swamps, and canals. **Range:** Common throughout the country, particularly in lowlands. **Regional Range:** China, Vietnam, Laos, Cambodia, Thailand, Peninsular and Eastern Malaysia, Singapore, and Indonesia (Kalimantan, Sumatra, Java and Sulawesi).

Sunbeam Snakes Family Xenopeltidae

The family is characterized by a cylindrical body. The tail is short. A single species is represented in Thailand.

170 Sunbeam Snake *Xenopeltis unicolor*

Body Length: Up to 96.4 cm.

A medium-sized snake, well-adapted for burrowing. The much-flattened head is wedge-shaped, and is scarcely distinct from the neck. The scales on the cylindrical body are very smooth. The eyes are small. The body is iridescent purple or chocolate-brown above. The ventral surface is white. There are 15 rows of scales throughout the body. **Behaviour:** Nocturnal and semi-burrowing habit. Found under logs and in debris. Diet consists of rodents, birds, lizards, and frogs. **Habitat:** Found in all lowland forest types and agricultural areas, and in forests up to over 1,000 m. **Range:** Widespread throughout Thailand. **Regional Range:** India (Andamans and Nicobar Islands), Myanmar, China, Vietnam, Laos, Cambodia, Thailand, Peninsular and Eastern Malaysia, Singapore, Brunei, Indonesia, and Philippines (Palawan).

170

Pythons Family Pythonidae

The family comprises the largest of the living snakes. The head is always distinct from the body. Their skeletons have the rudiments of a pelvic girdle. All of them have enlarged scales and sensory pits on their labials. The upper and lower jaws have teeth.

Pythons Genus *Python*

Three species, all from the genus *Python*, occur in Thailand. They are primarily terrestrial. They are often found near water bodies. Their diet consists of mammals and birds, and sometimes reptiles. All are oviparous.

KEY TO THE THAI *PYTHON*

1 a	Rostral and four upper labials on each side bearing pits; ventrals number from about 300 to 332; sub-caudals number 75-102; very large elongate snakes reaching a length of 8-9 m	*P. reticulatus*
1 b	Rostral and upper two labials on each side with pits	**2**
2 a	Ventrals number 245-275; sub-caudals number 58-75; large elongate snake reaching a length of 5-6 m	*P. molurus*
2 b	Ventrals number 160-175; sub-caudals number 28-32; a much smaller, shorter, thicker snake reaching to a length of about 2.5 m	*P. brongersmai*

Key modified from Taylor (1965).

171 **Reticulated Python** *Python reticulatus*

Body Length: Up to 900 cm.

The world's largest snake, in terms of length. The top of the head is uniform brown turning to light grey or dirty white on the supralabials. A narrow black line begins on the snout and runs over the back of the head to the nuchal region; this feature is absent in Burmese Python. A second narrow black line passes from each eye to the back of the head. On each side of the body there is a characteristic zigzag black stripe, which connects together across the back, enclosing a series of virtually oval brown spots. Below the zigzag lines are a series of ivory to yellowish spots. The rostral bears two sensory pits, as well as four supralabials on each side that each bear a single pit in which there are heat-sensing nerves to help detect warm-blooded prey; in the other Thai *Python* species there are only two supralabial pits. There are 70-80 scales around the middle of the body. The ventral scales number from 300 to 332, with the scales on the underside of the tail—the sub-caudals—numbering from 75 to 102. **Behaviour:** Nocturnal and usually terrestrial, though capable of climbing. It preys on mammals and birds. **Habitat:** Found in riverine, semi-evergreen, and evergreen forest.

171

Range: Found throughout the country. **Regional Range:** India, Bangladesh, Myanmar, Vietnem, Laos, Cambodia, Thailand, Peninsular and Eastern Malaysia, Singapore, Brunei, Indonesia, and Philippines. **International Conservation Status:** CITES: Appendix II.

---

172 Burmese Python *Python molurus*

Body Length: Up to 650 cm.

This species is a very large snake. The rostral and only 2 upper labials bear pits. The arrow-shaped mark on the head is diagnostic. The body has a ground colour of yellowish-brown or greyish-brown. On the dorsal surface it has large quadrangular, often somewhat elongated, reddish-brown, black-edged spots. On each side is a series of smaller spots with light centers. The girth is proportionately larger than that of the Reticulated Python, with scale-rows around the body numbering 61-75. The ventral scales number between 245 and 275. The sub-caudal scales number 58-75. **Behaviour:** Nocturnal and largely terrestrial. An ambushing constrictor. **Habitat:** Found in all lowland forest types and agricultural land. **Range:** Found throughout Thailand north of the Isthmus of Kra. Not presently recorded south of Surat Thani. **Regional Range:** Pakistan, India, Nepal, Bangladesh, Myanmar, China, Vietnam, Laos, Cambodia, Thailand, Peninsular and Eastern Malaysia, Brunei, and Indonesia. **International Conservation Status:** IUCN: Lower Risk—Near Threatened; CITES: Appendix II.

172

---

173 Blood Python *Python brongersmai*

Body Length: Up to 250 cm.

A much smaller and stouter snake than the other two species of *Python*. The head is proportionately small, short and lavender-coloured with light lines on labial edges. The rostral and the only two upper labials bear relatively deep transverse pits for sensing prey. The upper part of the dorsum is reddish or reddish-orange, with a row of cream- or fawn-coloured spots. The ventral scales are cream-coloured and number 160-175. The sub-caudal scales number 28-32. **Behaviour:** Nocturnal and terrestrial. Diet consists of mammals and birds. **Habitat:** Occurs in close association with rivers, streams, and other water bodies. **Range:** Occurs throughout peninsular Thailand, as far north as Petchaburi Province. **Regional Range:** Vietnam, Thailand, Peninsular Malaysia, Singapore, and Indonesia (Sumatra). **International Conservation Status:** CITES: Appendix II.

173

File Snakes Family Acrochordidae

This is a small family containing three heavy-bodied snakes with loose folded skin and granular scales. They are aquatic and thought to be nocturnal. They feed on fish and eels. File snakes lack ventrals and sub-caudals.

File Snakes Genus *Acrochordus*

In Thailand, 2 species occur.

KEY TO THE THAI *ACROCHORDUS*

1 a	Nostrils terminal directed forward; 130-150 scales around body; no distinct ventral median fold; no transverse bands; usually some longitudinal striping	*A. javanicus*
b	Nostrils dorsal, directed upward; approximately 100 scales around body; a well-defined ventral median fold; a banded colour pattern	*A. granulatus*

Key modified from Taylor (1965).

174 Javan File Snake *Acrochordus javanicus*

Body Length: Up to 230.4 cm.

A large extremely stout-bodied snake reaching 2.5 m long and a wide girth of up to 30 cm. The dorsal surface is generally dark olive, or grey. Sides and venter are pale yellow; the sides have a broken black stripe, which appear like blotches, running along the length of the body. The head has darker markings, and is indistinct from the body. The snout is very blunt, with nostrils pointing forward and upward. From 130 to 150 scales around the body. **Behaviour:** Nocturnal. Aquatic, feeding largely on fish. **Habitat:** Found in freshwater bodies, including rivers, canals, and peat swamp forest, as well as estuaries. **Range:** Found in central Thailand and peninsular Thailand, as well as the Gulf of Thailand. **Regional Range:** Vietnam, Laos, Cambodia, Thailand, Peninsular and Eastern Malaysia, Singapore, Brunei, and Indonesia.

174

175 Banded File Snake *Acrochordus granulatus*

Body Length: Up to 64.9 cm.

Large-girthed snake with a dirty-white body with about 52 black bands; these bands blend together on the back and tend to form a zigzag pattern along the posterior part of the body. The tail has a similar colour pattern, with 13 black bands. The nostrils are dorsal, pointing upward. A well-defined median ventral fold. Approximately 100 scales around body. **Behaviour:** Nocturnal. Aquatic, feeding on fish. Quiet and inoffensive. **Habitat:** Found in estuaries, mangrove, and coastal mudflats. **Range:** Found in coastal areas in the Gulf of Thailand and the Andaman Sea. **Regional Range:** Pakistan, India, Sri Lanka, Bangladesh, Myanmar, China, Vietnam,

175

Cambodia, Thailand, Peninsular and Eastern Malaysia, Singapore, Brunei, Indonesia, Philippines, Papua New Guinea, and Australia.

Colubrid Snakes Family Colubridae

The family contains a large proportion of the world's snake species, the majority of which are harmless. Dentition is the main characteristic used to distinguish this group from other families. Generally, colubrids have aglyphous teeth with no grooves or canals, or opisthoglyphous teeth with grooved fangs in the posterior upper jaw, being rear-fanged snakes. Their ventral scales are always of a different size to the scales on the upper body. Pit organs or other thermal sensors on the snout are absent. In Thailand they are represented by 7 genera.

Blunt-Headed Tree Snake Genus *Aplopeltura*

This is a monotypic genus. It has a head strongly distinct from the neck. The eye is large with a vertical pupil. The body is strongly compressed with a long prehensile tail. *Aplopeltura boa* occurs in Thailand.

176 Blunt-headed Tree Snake *Aplopeltura boa*

Body Length: Up to 75.0 cm.

Slender, compressed snake with a thickened head and blunt snout, reaching a length of 75 cm. It has large, bulging eyes. The distinct head has a conspicuous creamy white spot on the side of the face. It also has a vertical black mark running from the eye to the lip. The body is greyish-brown above, with indefinite black markings. There are irregular black blotches on lower sides. The scales are in 13 rows. **Behaviour:** Nocturnal. Arboreal, found in understorey bushes and undergrowth. Diet consists of slugs, snails, and lizards. **Habitat:** Inhabits lowland forests, in peninsular monsoonal evergreen forest and Malayan mixed dipterocarp forest **Range:** Found in peninsular Thailand. Recorded in Chumphon, Nakhon Si Thammarat, and Pattani provinces. **Regional Range:** Myanmar, Thailand, Peninsular and Eastern Malaysia, Brunei, Indonesia, and Philippines.

176

Slug-Eating Snakes Genus *Asthenodipsas*

Members of this genus have a single pre–chin shield anterior to the two pairs of chin shields. In the very closely related *Pareas* genus, there are no chin shields anterior to the 3 pairs of chin shields. They are nocturnal, terrestrial, and/or semi-arboreal, and have specialized diets of slugs and snails. All species lay eggs. Two species occur in Thailand.

KEY TO THE THAI *ASTHENODIPSAS*

1 a Six supralabials, with 3rd and 4th (sometimes 5th also) bordering orbit; 3rd infralabials forming a median structure with each other — *A. laevis*

b Seven supralabials, with 3rd and 4th bordering orbit; 2nd or 3rd infralabial joining to form a median structure — *A. malacannus*

Key prepared by Chan-ard, Parr and Nabithabhata.

177 Smooth Slug-eating Snake *Asthenodipsas laevis*

Body Length: Up to 54.5 cm.

A slender slug-eating snake with a laterally compressed body. The brown rounded head is very distinct from the slender body. It has large eyes with vertical pupils. The upper body is also mid-brown to dark brown, with irregular blackish crossbands that extend down to the venter. The sides are dull orange. The venter is cream or yellowish with brown spots. It can be identified by the combination of asymmetrical chin shields and 5-6 supralabials. Scales are in 15 rows throughout. **Behaviour:** Nocturnal. Semi-arboreal and terrestrial. Has a diet of slugs and snails. **Habitat:** Found in peninsular monsoonal evergreen forest. **Range:** Known from individuals in Nakhon Si Thammarat and Narathiwat provinces. **Regional Range:** Laos, Thailand, Peninsular and Eastern Malaysia, and Indonesia.

177

178 Malaccan Slug-eating Snake *Asthenodipsas malaccanus*

Body Length: Up to 44.0 cm.

A small snake with a laterally compressed body, reaching almost half a metre in length. The short, roundish head is sometimes whitish. The lower part of the neck and forebody is black. The dorsal surface is pale brown or mid-brown, with irregular brownish-grey crossbars. The venter is white or yellowish-cream, sometimes speckled with brown. It has 7-8 supralabials. **Behaviour:** Nocturnal. Terrestrial and semi-arboreal in forest undergrowth. Eats slugs and snails. **Habitat:** Found in lowlands in Malayan mixed dipterocarp forest. **Range:** Restricted to extreme south. Recorded only in Yala and Narathiwat provinces. **Regional Range:** Thailand, Peninsular and Eastern Malaysia, and Indonesia (Kalimantan and Sumatra).

178

Slug-Eating Snakes Genus *Pareas*

Members of this genus have very short, thick heads which are distinct from the neck. The eye is large or very large, with a vertical pupil. In arboreal forms, the body is much compressed. The tail is long, slender, and prehensile. In this genus, there are no chin shields anterior to the three pairs of chin shields while members of the genus *Asthenodipsas* have a single pre–chin shield anterior to two pairs of chin shields. The scales are in 15 rows throughout the body. They are nocturnal, terrestrial and/or semi-arboreal, oviparous, and have specialized diets of slugs and snails. Four species are represented in Thailand.

KEY TO THE THAI PAREAS

1 a	Vertebral scales not enlarged; body not strongly depressed; labials separated from orbit by sub-ocular; terrestrial	**2**
b	Vertebral scales of median row (sometimes adjoining rows also) more or less enlarged; body strongly depressed; arboreal	**3**

2 a Scales smooth, usually with, or without an interrupted neck band; ventrals less numerous — *P. margaritophorus*

b Scales keeled, usually with an irregular neck band; ventrals more numerous — *P. macularius*

3 a A single elongate sub-ocular (sometimes fused to postocular) — *P. hamptoni*

b Two to eight sub-oculars — *P. carinatus*

Key modified from Taylor (1965).

179 Keeled Slug-eating Snake *Pareas carinatus*

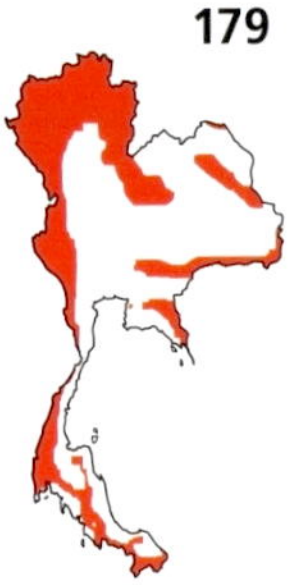

Body Length: Up to 48.8 cm.

A small slug-eating snake with a laterally compressed body and a large distinct head. The top of the head is light brown. The rostral, supralabials, and the chin are ivory yellow. A more-or-less continuous black line connects the back of the eye along the temporal region to the back of the neck. The labials are separated from the eye by between two and eight sub-oculars. The upper body is generally light-brown or olive-brown with up to 71 indistinct transverse darker marks crossing the body. The venter is pale-brown or yellow. **Behaviour:** Nocturnal. Arboreal, preferring the understorey. Diet consists of slugs and snails. **Habitat:** Found in dry evergreen forest, as well as southeastern and peninsular monsoonal evergreen forests. **Range:** Found in forested areas throughout the country. **Regional Range:** Myanmar, China, Vietnam, Laos, Cambodia, Thailand, Peninsular and Eastern Malaysia, and Indonesia.

179

180 White-spotted Slug-eating Snake *Pareas margaritophorus*

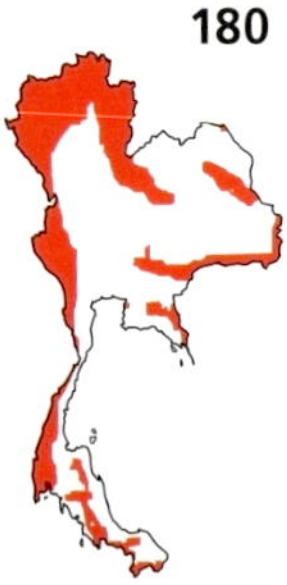

Body Length: Up to 32.0 cm.

A small-sized, laterally compressed slug-eating snake, barely reaching over 30 cm in length. The thickened head, which is distinct from the neck, is lavender-blue. It usually has a conspicuous creamy-white nuchal collar, about four cells wide, which contrasts strongly with a dark band immediately behind; in some individuals this feature is absent. The body is generally lavender-grey or light brown, with irregular black crossbars bordered with white spots. The dorsal scales are smooth; in the similar Black-spotted Slug Snake (*macularius*) the scales are keeled. The ventral scales number 138-159. **Behaviour:** Nocturnal. Both terrestrial and arboreal. Feeds on molluscs. **Habitat:** Found in understorey vegetation in evergreen forest. **Range:** Found in forested areas throughout the country. **Regional Range:** Myanmar, China, Vietnam, Laos, Cambodia, Thailand, and Peninsular Malaysia.

180

---

181 Black-spotted Slug-eating Snake *Pareas macularius*

Body Length: Up to 21.0 cm.

This small species closely resembles the White-spotted Slug Snake; it has a slightly more laterally compressed body. It has a thick rounded head, distinct from the body which is speckled brown, and a pair of black spots behind the parietals. It usually has an irregular neck collar, which is creamy white. The body is bluish-grey with approximately 60 transverse rows of scales with black dots bordered with white; the rows are irregular and often incomplete. The 5-7 median dorsals are keeled, with a somewhat larger series on the ventrals and sub-caudals. **Behaviour:** Nocturnal. Both terrestrial and arboreal. Diet consists of slugs and snails. **Habitat:** Found in

181

understorey vegetation in dry evergreen forest and sub-montane forest. **Range:** Found in Chiang Mai and Loei provinces. **Regional Range:** India, Bangladesh, Myanmar, China, Vietnam, Laos, and Thailand.

182 Hampton's Slug-eating Snake *Pareas hamptoni*

Body Length: Up to 70.5 cm.

A large slug-snake with a strongly laterally compressed body. The head, which is sharply distinct from the neck, is marked with dark grey on top and forehead. A black line passes from the eye to the nape. The body is brown above with vertical blackish bars on sides which may extend across the back. The vertebral scales are enlarged. The underside is yellowish, dotted with brown. The ventrals number 180-194. **Behaviour:** Nocturnal. Terrestrial and arboreal, being found close to forest streams. **Habitat:** Presently recorded in southeastern monsoonal evergreen forest and peat swamp forest, though may occur in other forest types. **Range:** Recorded in Chanthaburi Province in southeastern Thailand and Bang Nara, Narathiwat Province in peninsular Thailand. Seemingly rare. **Regional Range:** Myanmar, China, Vietnam, Laos, Cambodia, and Thailand.

182

Dragon Snake Genus *Xenodermus*

This genus comprises a single species, with strange body skin, and is sometimes referred to as a "dragon" snake. The head is covered in small granular scales bearing keels rather than head shields; the scales in the mid-dorsal region are varied, with a median series of somewhat enlarged scales alternating with smaller scales. It has a slender, feebly compressed body. The pupil is vertical. The nostril is a single nasal which is pointed forward. One species, *Xenodermus javanicus*, is represented in Thailand.

183 Rough-bodied Draogon Snake *Xenodermus javanicus*

Body Length: Up to 43.3 cm.

A small slender-bodied snake, with a slate grey head, a black-tipped snout and whitish marks on the nasals and rostral. The body is only very slightly compressed; the dorsum is unpatterned slate-grey, with light-grey on the sides. It has three rows of large keeled scales on the dorsum which are dark grey. The venter is grey, with each ventral having a grey-brown mark along its anterior edge. The sub-caudals are barred with brown. **Behaviour:** Noctural and terrestrial, although reported to burrow. Found in wet earth or rotting logs or close to streams. Feeds on amphibians, particularly frogs. **Habitat:** Found in Malayan mixed dipterocarp forest. **Range:** Southernmost provinces of peninsular Thailand. Known from only two specimens found in Yala Province. **Regional Range:** Myanmar, Thailand, Peninsular and Eastern Malaysia, Brunei, and Indonesia.

183

Racers Genus *Coelognathus*

This genus comprises moderately large ratsnakes. They normally have three supralabials in contact with the eye, and usually lack an anterior sub-ocular. The anal plate is usually single. Most are either diurnal or crepuscular, being largely terrestrial in nature. They often have a varied diet, preying upon frogs, lizards, birds, and small mammals. Two species occur in Thailand.

KEY TO THE THAI *COELOGNATHUS*

1	**a**	A single narrow black transverse occipital bar or band with black lines running forward from bar to eye; head coppery brown	*C. radiatus*
	b	No transverse bars; scales keeled; a median light yellowish-brown stripe flanked by two narrow black stripes anteriorly; a series of ventrolateral spots or ocelli on anterior third of body	*C. flavolineatus*

Key modified from Taylor (1965).

184 Yellow-striped Racer *Coelognathus flavolineatus*

Body Length: Up to 141.9 cm.

A large snake reaching approximately 1.5 m in length. Olive-brown elongate head is scarcely distinct from the body; sides of head are olive-brown. Supralabials are ivory-yellow. Two narrow dark postocular stripes radiate backward from the eye, with another narrow stripe across nape. Dorsal colour is brownish-grey or brownish-yellow, with a pair of sometimes bold dorso-lateral black lines running along the anterior part of the body, between which is a narrow yellow stripe. The flanks have short black stripes or blotches. The venter is pale yellow, being somewhat darker posteriorly. Scales in 19 rows at mid-body. **Behaviour:** Terrestrial and arboreal habits. Diet includes rodents, birds, lizards, and frogs. **Habitat:** Inhabits dry evergreen forest, peninsular monsoonal evergreen forest, and Malayan mixed dipterocarp forest. **Range:** Found in northeastern and peninsular Thailand. Recorded in Nakhon Ratchasima Province and southernmost provinces. **Regional Range:** India (Andaman and Nicobar Islands), Myanmar, Vietnam, Thailand, Peninsular and Eastern Malaysia, Singapore, Brunei, and Indonesia.

184

185 Copperhead Racer *Coelognathus radiatus*

Body Length: Up to 144.0 cm.

Head is coppery brown, with a narrow black line radiating from the eyes downward, a second radiating backward and a third radiating at an angle diagonally in-between. The transverse line meets a thin black collar line. Body, which is highly compressed in the area immediately behind the head and neck, is yellowish-brown or grey, with two broad black stripes running down the dorsum on each side. These black stripes gradually narrow, turn brown, and fade out. The tail is light-brown. Scales in 19 rows at mid-body. **Behaviour:** Diurnal and largely terrestrial habit. Diet consists of frogs, birds, and rats. When disturbed, the snake inflates the area behind the neck. **Habitat:** Found in the lowlands, particularly grasslands, open forests, and paddy fields, as well as forests up to 1,400 m. **Range:** Present throughout

Thailand. **Regional Range:** India, Nepal, Bangladesh, Myanmar, China, Vietnam, Laos, Cambodia, Thailand, Peninsular and Eastern Malaysia, Singapore, and Indonesia.

185

Racer Genus *Oreocryptophis*

This genus comprises a single species. It is either crepuscular or nocturnal, being largely terrestrial in nature. It preys upon warm-blooded animals. One species occurs in Thailand.

186 Red Mountain Racer *Oreocryptophis porphyraceus*

Body Length: Up to 110.0 cm.

Strikingly coloured snake, having a dull reddish-orange to magenta dorsum with a series of 14–16 dark bluish-black bands on the body and tail. Two narrow black lines extend straight forward to the eyes from the first band on the head, as well as a median black line passing from the snout to the end of parietals. Venter is unpatterned cream or yellow. Scales are smooth in 19 rows. **Behaviour:** Crepuscular or nocturnal. Mainly terrestrial, but may also have semi-arboreal habit, being found in low understorey vegetation. Diet consists of small mammals. **Habitat:** Found in sub-montane and

186

upper montane forests. **Range:** Found in mountains in northern and north-eastern (west) Thailand. Recorded in Chiang Mai, Loei, and Chaiyaphum provinces. **Regional Range:** India, Nepal, Myanmar, China, Vietnam, Laos, Thailand, Peninsular Malaysia, and Indonesia (Sumatra).

Racers Genus *Orthriophis*

This genus comprises a small group of ratsnakes that were formerly assigned to the genus *Elaphe*. Based on DNA evidence, a new genus was created in 2002. The name derives from the Greek *orthros* for "dawn" and *ophis* for "snake", referring to the snakes' high level of activity in the early morning. They prey upon warm-blooded animals, particularly bats. One species occurs in Thailand.

187 Yellow Black-tailed Racer *Orthriophis taeniurus*

Body Length: Up to 188.2 cm.

Two sub-species of this Racer occur in Thailand. *O. t. ridleyi* has a dull slate head above with a black stripe through the loreal region. Supralabials are yellow. The anterior half of dorsum is yellowish-fawn; middle half of the body and sides become darker, changing from greyish-slate to black. A median yellow stripe begins on the dorsum at about the middle of the body and becomes more intense until the tip of the tail. The venter is yellowish. *O.t.helfenbergeri* has a light-brown head and neck, with a black stripe running through the eye. Anterior part of back covered in large black butterfly-shaped spots with smaller diamond-shaped spots on sides. The posterior part of the back has a grey vertebral stripe bordered by a broad black stripe on each side. **Behaviour:** Diurnal and nocturnal. Often lives

187

in caves, feeding on bats and swiftlets. **Habitat:** Inhabits lowland and sub-montane forest, often associated with limestone. Cave dwelling. **Range:** *O.t.ridleyi* restricted to southernmost provinces of peninsular Thailand. *O.t.helfenbergeri* found in northern and western Thailand. **Regional Range:** India, Myanmar, China, Vietnam, Laos, Thailand, Peninsular and Eastern Malaysia, and Indonesia (Kalimantan and Sumatra).

Racers Genus *Gonyosoma*

Members of this genus are large and slender. The body is slightly compressed laterally. The head is slender. The pupil is round. They have 19-27 dorsal scales at mid-body. Three species occur in Thailand.

KEY TO THE THAI *GONYSOMA*

1 a	Uniform green on body	**2**
b	Light or dark buff marked all over by scattered irregular blotches of dark brown or black; venter uniform whitish or with spots similar to those on dorsum; two supralabials enter eye, third touches it at a point	*G. floweri*
2 a	Interstitial skin with black and white reticulations; scale-rows at mid-body 23 (25)	*G. oxycephalum*
b	Interstitial skin with blue reticulations; 19 scale-rows at mid-body	*G. prasina*

Key modified from Taylor (1965).

188 **Red-tailed Racer** *Gonyosoma oxycephalum*

Body Length: Up to 210.0 cm.

A large, emerald green–coloured snake exceeding 2 m in length, with a tail that is characteristically either light-chestnut or buffy-red. It has a long, slightly pointed snout, being 3 times as long as eye. The green head sometimes has a sharply defined black line which passes through the eye. Immediately below this line the chin is brilliant pale green. It has a blue tongue. The ventrals are light greenish-yellow, and are slightly keeled. **Behaviour:** Arboreal and diurnal. Preys upon small rodents, birds, and bats. **Habitat:** Inhabits primary and secondary evergreen forest, including sub-montane forest, mangrove, swamps, and marshes, scrubland, and plantations. **Range:** Found in forested areas throughout the country. **Regional Range:** India (Andamans and Nicobar Islands), Myanmar, Vietnam, Laos, Cambodia, Thailand, Peninsular and Eastern Malaysia, Singapore, Brunei, Indonesia, and the Philippines.

188

189 Green Mountain Racer *Gonyosoma prasina*

Body Length: Up to 120.0 cm.

Dorsal surface and sides of body are green or turquoise without any markings. The tip of the tail is brown. The top of the head is dark-green. The snout, chin and throat are very pale green. Faint dark postocular stripe is present. The tongue is reddish brown. Ventrals are ivory-edged with yellow and turquoise-blue. **Behaviour:** Diurnal and arboreal. Diet includes birds, mammals, and lizards, which are killed by constriction. **Habitat:** Inhabits sub-montane forest. Found near streams. **Range:** Occurs in northern and northeastern (west) Thailand. **Regional Range:** India, Bhutan, Myanmar, China, Vietnam, Laos, Thailand, and Peninsular Malaysia.

189

190 Flower's Racer *Gonyosoma floweri*

Body Length: Up to 180.0 cm.

A long snake, being almost 2 m long. Light- or dark-buff head and body, the latter covered by irregular blotches of dark brown or black. Venter is uniformly white. Tail is spotted. **Behaviour:** Arboreal. **Habitat:** Found in peninsular monsoonal evergreen forest and Malayan mixed dipterocarp forest. **Range:** Found in southern provinces; northern limit is Trang Province. **Regional Range:** Endemic to Thailand.

190

Rat Snakes Genus *Ptyas*

This genus comprises moderately large to large, slender snakes. They are diurnal and very agile, capable of moving at great speed. The diet is highly varied; they feed on rodents, birds, snakes, lizards, and frogs. They are not constrictors, but most usually swallow prey live. They have slightly compressed bodies. The head is long and distinct from the body. The eye is large and the pupil round. A small sub-ocular scale is often present. The posterior maxillary teeth are only slightly enlarged. The ventrals are rounded. The tail is long. The sub-caudals are in two rows. Four species occur in Thailand.

KEY TO THE THAI *PTYAS*

1 a	Scale-rows in odd numbers	**2**
b	Scale-rows in even numbers	**3**
2 a	Scales in 15 rows at mid-body; ventrals less numerous, numbering 160-187; sub-caudals number 120-147; young with white bands; adults showing lineation in posterior part of body	*P. korros*
b	Scales in 17 rows at mid-body; ventrals 197; sub-caudals 117; darker bands at least in posterior part of the body, that may be somewhat reticulated or not; labials and neck scales with black sutures	*P. mucosa*

3 a The two dorsal-most rows of scales keeled; a bold black-and-yellow checquered pattern posteriorly and on the tail *P. carinata*

b All the scales smooth; tail and posterior part of the body brown with a pair of black lateral stripes sharply marked off from the pale underparts; labial without black sutures *P. fusca*

Key modified from Taylor (1965); Tweedie (1983).

191 Indochinese Rat Snake *Ptyas korros*

Body Length: Up to 140.2 cm.

Large snake reported to reach at least 2 m, being often mistaken for the King Cobra. The head and anterior part of the body is brownish grey or grey. Posterior part of body darkens to nearly black. Sides and tip of snout are whitish. Labials, chin, and ventral surface are brownish-cream. Posterior part of sub-caudal region is white. Number of scale-rows is never in even numbers. **Behaviour:** Mainly terrestrial habit. Often seen in bushes or low trees particularly near water. Feeds on frogs, rats, and other small rodents. **Habitat:** Found in both lowland and montane forests. **Range:** Present in forests throughout Thailand. **Regional Range:** India, Myanmar, China, Vietnam, Laos, Cambodia, Thailand, Peninsular and Eastern Malaysia, Singapore, and Indonesia.

191

192 Common Rat Snake *Ptyas mucosa*

Body Length: Up to 175.3 cm.

Very large slender snake often mistaken for the King Cobra. Head is dark brown above, with yellowish snout. Labials yellowish with black centres. Body is amber to brown, with the posterior three-quarters of body having narrow irregular dark bands. Venter is light yellowish-brown anteriorly, becoming more yellowish posteriorly. Sub-caudals are ivory-yellow. **Behaviour:** Both terrestrial and arboreal habit. Feeds on rats, other small rodents, and frogs. **Habitat:** Found in mixed deciduous forest as well as agricultural areas. Found in rice paddies. **Range:** Present throughout Thailand. **Regional Range:** Iran, Afghanistan, Pakistan, Turkemenistan, India, Nepal, Sri Lanka, Bangladesh, Myanmar, China, Vietnam, Laos, Cambodia, Thailand, and Indonesia (Sumatra and Java). **International Conservation Status:** CITES: Appendix II.

192

---

193 Keeled Rat Snake *Ptyas carinata*

Body Length: Up to 300.0 cm.

A very large slender snake, reaching 3 m in length. Anterior portion of the body is olive-brown to nearly black; the posterior part is chequered black and green. The tail is black with yellow spots. The venter is cream, turning grey or black posteriorly. Scale-rows in even numbers. **Behaviour:** Diurnal. Mainly terrestrial habit, but sometimes arboreal. Feeds on frogs, rats, and other small rodents. **Habitat:** Found in all lowland forest types and agricultural areas. **Range:** Present throughout Thailand. **Regional Range:** Myanmar, China, Vietnam, Thailand, Peninsular and Eastern Malaysia, Singapore, Indonesia, and the Philippines (Palawan).

193

---

194 White-bellied Rat Snake *Ptyas fusca*

Body Length: Up to 290.0 cm.

The brown head is long and distinct. The slender body is brown above, with a red stripe sometimes present along the vertebral ridge. The posterior part of the body and the tail have a black lateral stripe that gradually widens; the tail may be entirely black. The contrast between this black stripe and the venter which is whitish-yellow is striking. **Behaviour:** Terrestrial habit (usually found sleeping on tree branches at night like all species of *Ptyas*). Preys on birds, rats, other small rodents, and lizards. Oviparous. **Habitat:** Found in evergreen forest and rubber plantations. **Range:** Rare. Found in Peninsular Thailand. Recorded in Phang-nga, Krabi, and Nakhon Si Thammarat provinces. **Regional Range:** Thailand, Peninsular and Eastern Malaysia, Singapore, Brunei, and Indonesia.

194

Brown Snakes Genus *Xenelaphis*

This genus comprises terrestrial snakes. The head is distinct from the neck. It has a cylindrical body. The eye is fairly large and the pupil is round. The dorsals are smooth. The vertebrals are slightly enlarged and hexagonal. The tail is long. The sub-caudals are divided in two rows. There are 17 scale-rows at mid-body. One species is represented in Thailand.

195 **Malayan Brown Snake** *Xenelaphis hexagonotus*

Body Length: Up to 198.0 cm.

A large robust snake, sometimes reaching 2 m. The head is dark-olive to brown with a pair of blackish markings surrounded by tiny yellow dots. Body is brownish-grey or dark green with an indistinct median row of tiny black dots and lighter dots running along the vertebral ridge. All ventral scales have black spots on the outer edges; anterior ventrals also have a pair of black dots in the centre. The young are light brown with a series of black bands crossing the body; these bands fade with age. **Behaviour:** Diurnal and terrestrial. Feeds on rats, mice, and birds. **Habitat:** Occurs in peninsular monsoonal evergreen forest, mixed dipterocarp, mangrove, and peat swamp forests. **Range:** Found in peninsular Thailand, northward to Surat Thani Province. **Regional Range:** Vietnam, Thailand, Peninsular and Eastern Malaysia, Singapore, Brunei, and Indonesia (Kalimantan, Sumatra, and Java).

195

Ringnecks Genus *Gongylosoma*

Representatives of this genus are small snakes. They are terrestrial, living on the forest floor in debris. In Thai species, the head is either not distinct or scarcely distinct from the neck. The body is cylindrical, and the tail is long. The sub-caudals are paired. There are 13 scale-rows around the body. Three species occur in Thailand.

KEY TO THE THAI *GONGYLOSOMA*

1 a A white or yellow chevron-shaped collar on the nape; anterior part of the body with five orange stripes; ventrals number 110-122; sub-caudals 77–103 — *G. longicauda*

b Occiput with a crossband or ringed collar on the nape; anterior part of body with crossbars or transverse spots — **2**

2 a A narrow interrupted band across occiput, bordered with black, followed by a darker saddle-shaped blotch; ventrals number 117–145; sub-caudals 89–103 — *G. scriptum*

b Several dark bands around anterior part of body; eight supralabials; ventrals number 115–137; sub-caudals 58–75 — *G. baliodeirum*

Key modified from Taylor (1965); Tweedie (1983).

196 Common Ringneck *Gongylosoma scriptum*

Body Length: Up to 35.5 cm.

Small snake, reaching almost 40 cm in length. The body is generally greyish-brown above, sometimes with the scales edges being darker brown or black, forming longitudinal stripes. The head is greyish to greyish-brown above. It usually has a broad, black-edged cream nuchal collar, preceding a dark neck band. The scales are in 13 rows throughout the body. The ventrals and sub-caudals are cream-coloured, numbering 117-145 and 89-103, respectively. **Behaviour:** Diurnal and terrestrial. Lives in the leaf litter and forest debris. **Habitat:** Occurs in sub-montane forest and peninsular

196

monsoonal evergreen forest. **Range:** Found in western, northeastern (west), and peninsular Thailand. **Regional Range:** Myanmar and Thailand.

197 Striped Ringneck *Gongylosoma baliodeirum*

Body Length: Up to 37.4 cm.

The uniform light-brown head is a little wider than the neck. It has 8-9 blackish bands around the neck and the anterior part of the body, separated by dull cream- or fawn-coloured bars. The posterior portion of the body is violet-brown, with each scale having a diamond-shaped, light lavender blue center. It has 13 scale-rows around the body. **Behaviour:** Lives in the leaf litter and in forest debris. **Habitat:** Found in southeastern monsoonal evergreen forest and Malayan mixed dipterocarp forest. **Range:** Restricted to Khao Soi Dao Wildlife Santuary in southeastern Thailand and extreme southern peninsular Thailand. **Regional Range:** Thailand, Peninsular and Eastern Malaysia, Singapore, Brunei, and Indonesia.

197

198 Long-tailed Ringneck *Gongylosoma longicauda*

Body Length: Up to 45 cm.

Dark brown above with five orange or yellow longitudinal stripes, clearly marked anteriorly, disappearing about a third of the way along the body, the dorso-lateral stripes widening and turning downward on the neck. A white or cream chevron-shaped collar on the neck. The top of the head is brown. A dark mark under the eye and the labial shields spotted with black. Underparts are red or yellow. **Behaviour:** Lives in the leaf litter and in forest debris. **Habitat:** Found in Malayan mixed dipterocarp forest. **Range:** Restricted to extreme peninsular Thailand. Recorded in Narathiwat Province. **Regional Range:** Thailand and Peninsular Malaysia.

198

Ringnecks Genus *Liopeltis*

Representatives of this genus are small snakes. They are terrestrial, living on the forest floor in debris. In Thai species the head is either not distinct or scarcely distinct from the neck. The body is cylindrical, and the tail is long. The sub-caudals are paired. There are 13-17 scale-rows around the body. One species occurs in Thailand.

199 **Three-coloured Ringneck** *Liopeltis tricolor*

Body Length: Up to 56.0 cm.

The largest Ringneck, reaching just over half a metre. The dorsal body coloration is greenish or olive. A conspicuous black stripe runs through the eye and onto the anterior part of the body. The pupils are round. The ventral surface is yellow. The ventrals number 140-187 and the sub-caudals number 103-147. It has 15 dorsal scale-rows at mid-body. **Behaviour:** Diurnal. Terrestrial species, living in the leaf litter. Feeds on insects and spiders.

199

Habitat: Occurs in Malayan mixed dipterocarp forest. Found up to 1200 m altitude. **Range:** Restricted to extreme southern Thailand. **Regional Range:** Thailand, Peninsular and Eastern Malaysia, Singapore, Brunei, Indonesia, and the Philippines (Palawan).

Kukri snakes Genus *Oligodon*

The Kukri snakes comprise rather short, medium-sized snakes with short, indistinct heads. Many species have a distinct chevron mark on the neck. They are called "Kukri" because of the shape of the fangs at the back of the mouth, which resemble the Kukri knife used by Gurkha soldiers. Kukri snakes, however, are non-venomous and quite harmless, though they may attempt to bite if provoked. Some species have highly variable markings and identification is not straightforward. They are terrestrial in habit and usually nocturnal. They tend to live under debris. All are oviparous. The scale-rows range from 13 to 23 throughout the body. Ten species are represented in Thailand.

KEY TO THE THAI OLIGODON

1 a	13 mid-body scale-rows; no loreal	*O. catenata*
b	15-23 mid-body scale-rows; loreal present	**2**
2 a	15 mid-body scale-rows, internasals and prefrontals fused, body much elongated; pale rings on a dark brown body	*O. jintakunei*
b	15-23 mid-body scale-rows, internasals and prefrontals distinct, body usually rather stout, striped or marked with dark crossbands on a lighter body	**3**
3 a	Dorsal scale-rows at mid-body number 15	*O. inornatus*
b	Dorsal scale-rows at mid-body number 17 or more	**4**
4 a	Mid-body scale-rows number 17	**5**
b	Mid-body scale-rows number 19 or more	**8**
5 a	Body with at least two longitudinal stripes	**6**
b	Body patternless or with rounded or transverse markings	**7**
6 a	Usually seven supralabials; pattern made of transverse elongated dark dorsal spots, separated by other dark spots, with one pale stripe on each flank	*O. barroni*
b	Eight supralabials; pattern entirely made of longitudinal stripes	*O. mouhoti*
7 a	More than 183 ventrals	*O. joynsoni*
b	Fewer than 186 ventrals	*O. cinereus*
8 a	Dorsum with longitudinal stripes, with or without another kind of pattern	*O. taeniatus*
b	Dorsum without longitudinal stripes, only rounded or transversal markings	**9**

9 a Usually 21 (sometimes 23) mid-body scale-rows; 160-190 ventrals; dorsal pattern either reticulated or with 13-18 transverse blotches separated by three or four irregular but wide dark crossbands, or with irregular crossbands *O. fasciolatus*

b Usually 19 (sometimes 21) mid-body scale-rows; 160-210 ventrals; dorsal pattern of wide dorsal blotches separated by three irregular, thin, and faint crossbands *O. purpurascens*

Key modified from Smith (1943); Taylor (1965).

200 Ashy Kukri Snake *Oligodon cinereus*

Body Length: Up to 63.2 cm.

This species may be distinguished from other medium-sized Kukri snakes by having 17 scale-rows around the mid-body. Two sub-species occur, both of which have eight supralabials. *O.c. swinhonis* is generally fawn-brown to grey on the dorsal surface. On the neck is a well-defined narrow dark brown chevron. It has a faint mark below the eye. The other sub-species, *O. c. multifasciatus*, is brownish-grey with irregular transverse black markings. This sub-species has a band running across the snout through the eyes and a diagonal mark above the angle of the mouth. In *O.c. swinhonis* the venter is yellowish-white, lacking ventral spots; in *O. c. multifasciatus* the venter is ivory yellow with a distinct double row of black spots. Both sub-species have 155-185 ventrals. **Behaviour:** Terrestrial and nocturnal. Diet includes insects and spiders. **Habitat:** Found in dry evergreen forest. **Range:** Recorded in three localities, namely Muang and Chiang Dao districts, Chiang Mai Province and Sakaerat, Nakhon Ratchasima Province. **Regional Range:** India, Bangladesh, Myanmar, China, Vietnam, Laos, Cambodia, Thailand, Peninsular and Eastern Malaysia.

200

201 Inornate Kukri Snake *Oligodon inornatus*

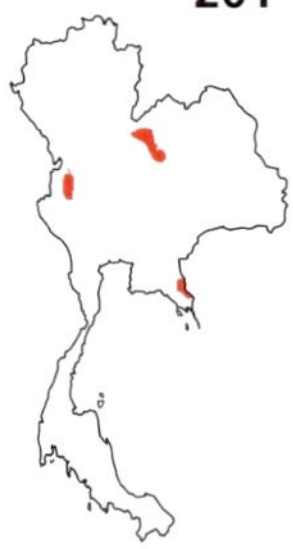

Body Length: Up to 58.0 cm.

A short, medium-sized snake. It is uniform pale brown or brown above, though may have a reticulated pattern in which black pigmentation is confined to the scale edges. The head is grey or dark brown, lacking any typical head pattern characteristic of the genus. The absence of head and body markings, which is distinctive for this Kukri species, explains the name 'inornate' or unicoloured. It has eight supralabials, and nine or more maxillary teeth. The ventral surface is usually uniformly white or yellowish, sometimes with occasional brown markings. The absence of spots on the white or yellowish-white sub-caudals separates the species from several other Kukri snakes. It has 15 rows around the mid-body. The anal is entire. **Behaviour:** Mainly terrestrial habit. Nocturnal. **Habitat:** Found in dry evergreen forest and southeastern monsoonal evergreen forest. **Range:** Recorded from western, northeastern (west), and southeastern Thailand. **Regional Range:** Cambodia and Thailand.

201

202 Joynson's Kukri Snake *Oligodon joynsoni*

Body Length: Up to 76.0 cm.

A medium-sized Kukri snake, measuring more than half a metre long. Head and body markings similar to Banded Kukri Snake, with a narrow dark brown chevron terminating on the nape. Dark purplish-brown above with ill-defined and broken crossbands, every alternate one enlarged. It lacks the longitudinal line markings found in a number of other Kukri snake species. It has an oblique brown temporal streak. The venter is red, heavily barred with black. The scale-rows are in 17 rows around the mid-body. The ventrals number 187-197, with 40-50 sub-caudals. By comparison, Ashy Kukri Snake has fewer than 186 ventrals. **Behaviour:** Terrestrial and nocturnal. **Habitat:** Found in lower and upper montane forests and dry evergreen forests. **Range:** Recorded in Khamphaeng Phet Province, northern Thailand and from Khao Yai National Park, northeastern Thailand. **Regional Range:** Vietnam, Laos, Cambodia, and Thailand.

202

203 Purple Kukri Snake *Oligodon purpurascens*

Body Length: Up to 22.2 cm.

A medium-sized Kukri snake. Distinctive dorsal patterning, being purplish-brown above with about 15 quadrangular dark brown or black saddle-like bands, often with dark borders, at regular intervals along the body, with a further four on the tail; the interspaces have 2-3 darker transverse lines. There is a dark band extending from the angle of the mouth through the eye to the forehead. It has 160-210 ventrals. It usually has 19, sometimes 21, scale-rows at mid-body. **Behaviour:** Nocturnal. Terrestrial. Very slow and sluggish, spending much of its time beneath rocks. Feeds on the eggs of lizards and frogs, as well as on frogs and tadpoles. **Habitat:** Found in peninsular monsoonal evergreen forest and Malayan mixed dipterocarp forest. **Range:** Found in peninsular Thailand. **Regional Range:** Thailand, Peninsular and Eastern Malaysia, Singapore, and Indonesia.

203

204 Blyth's Kukri Snake *Oligodon catenatus*

Body Length: Up to 60.7 cm.

One of the larger Kukri snake. Light brown above with a pair of dark-brown median stripes extending the entire length of the body to the tail, posteriorly separated by a light median stripe. It has a brown lateral stripe running along the side of the body until the vent. It also has 10 dorsal blotches on the back. On the back of the head is a heart-shaped chevron, with the apex of the chevron extending to the middle of the frontal. A second dark-brown mark passes obliquely from the eye to the base of the mouth. Another mark passes from the snout through the eye to the 5th supralabial. The chin and venter is ivory and immaculate. The tail is long, with four dorsal spots. It has more than 175 ventrals. The scale-rows number 21 at mid-body. **Behaviour:** Nocturnal and terrestrial. **Habitat:** Found in sub-montane forest. **Range:** Confined to western and northeast Thailand. **Regional Range:** India, Myanmar, and Thailand.

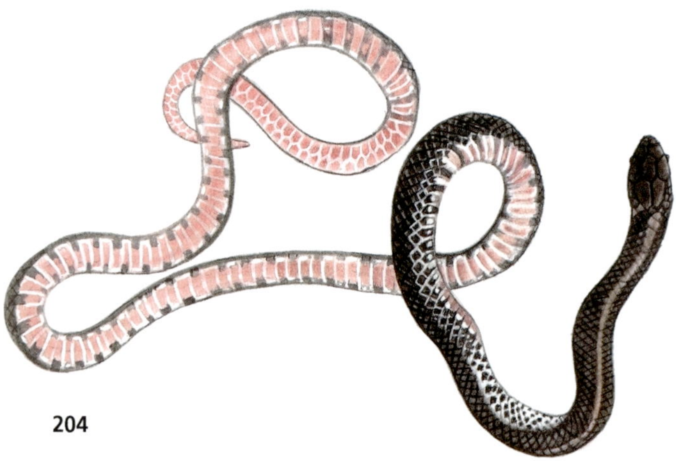

204

205 Jintakune's Kukri Snake *Oligodon jintakunei*

Body Length: Up to 44.8 cm.

A small-sized snake with a gracile and much-elongated body. The characteristic dorsal body patterning consists of 11 regularly spaced, narrow, whitish rings touching the ventrals on a dark-brown background. The head is clearly distinct from the neck, and shows the crescent-shaped lupus. The throat is white. The internasals and prefrontals are fused with one another on the side of the head. The eye is moderate and the pupil is rounded. The venter is white. It has 189 ventrals and 46 pairs of sub-caudals. It has 15 dorsal scale-rows at mid-body. **Behaviour:** Inhabits lowland forest. Other aspects of its natural history are unknown. **Habitat:** Found in peninsular monsoonal evergreen forest and Malayan mixed dipterocarp forest. **Range:** Recorded from Krabi Province in peninsular Thailand. **Regional Range:** Endemic to Thailand.

206 Banded Kukri Snake *Oligodon fasciolatus*

Body Length: Up to 80.7 cm.

A small Kukri snake. The head is tan, brown, or greyish-brown above, unmarked or with random markings. The dorsal body pattern is varied, being tan, brown, or grey-brown, either reticulated and dark-barred with 13-18 transverse blotches separated by 3-4 irregular (but wide) dark crossbands or with regular crossbands. The venter is white or yellow, sometimes with random brown dots. It has 160-190 ventrals. **Behaviour:** Nocturnal and terrestrial. **Habitat:** Found in dry dipterocarp forest and mixed deciduous forest. Also occurs in agricultural and urban areas. **Range:** Recorded from northern, northeastern, and southeastern Thailand. **Regional Range:** Vietnam, Laos, Cambodia, Thailand, and Peninsular Malaysia.

206

207 Striped Kukri Snake *Oligodon taeniatus*

Body Length: Up to 40.2 cm.

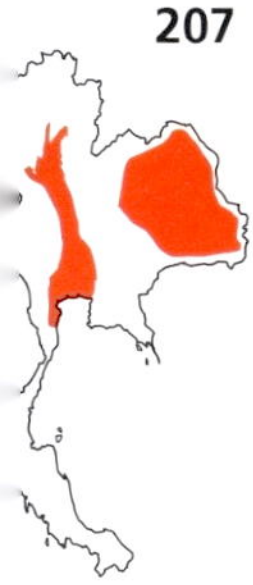

A medium-sized snake with a greyish-brown head and body. It has two dorsal blackish stripes, which are separated by a narrow yellowish-white stripe, and two lateral brown stripes. It lacks the transverse blotches or crossbands found in other Kukri snakes. The head pattern includes two oblique dark-brown markings and a brown arrowhead marking on the nape; the nuchal collar does not extend to the ventrals as in Cambodian Kukri Snake. The snout has a wide black mark. The chin and throat are dirty ivory. Lateral portion of venter are cream or ivory, graduating through pink and becoming red toward the tip. The dorsal scales are in 19 rows. **Behaviour:** Nocturnal. Mainly terrestrial habit. Diet consists of frogs and lizards and their eggs. **Habitat:** Found in dry dipterocarp forest, and also occurs in agricultural areas and urban areas. **Range:** Present in northeastern and central Thailand. Recorded in Bangkok. **Regional Range:** China, Vietnam, Laos, Cambodia, and Thailand.

207

208 Cambodian Kukri Snake *Oligodon mouhoti*

Body Length: Up to 33.9 cm.

A small Kukri snake. Grey in colour with two light-brown stripes extending along the vertebral column. The black nuchal collar extends to the lateral edges of the ventrals. There are two dark-brown spots on the dorsum of the tail, one at the base and one at the tip; this feature is absent in the similar Striped Kukri Snake. The ventral surface is pink, with 147-152 ventrals. The

208

dorsal scales are in 17 rows at mid-body. **Behaviour:** Terrestrial. **Habitat:** Found in agricultural areas near forests. **Range:** Recorded in southeastern and central Thailand. **Regional Range:** Vietnam, Laos, Cambodia, and Thailand.

209 Barron's Kukri Snake *Oligodon barroni*

Body Length: Up to 36.4 cm.

A medium-sized snake with a generally light brown body with 10-12 dark brown bars on the body and tail. It also has faint traces of two dorsal stripes of darker brown and a still fainter lateral stripe. Head has a dark brown transverse crescent. On the nape is a heart-shaped, dark brown nuchal spot. Venter is pink or red with two rows of quadrangular markings extending onto the sub-caudals. **Behaviour:** Nocturnal and terrestrial. **Habitat:** Found in lowlands up to 300 m. **Range:** Occurs in southeastern and central Thailand. **Regional Range:** Vietnam, Laos, Cambodia, and Thailand.

209

Reed Snakes Genus *Calamaria*

This genus comprises small snakes with heads indistinct from their round bodies. They are fossorial in nature, living in the leaf litter, where they forage for food at night. Their diet consists of worms. The eyes are distinct. They have short tails. The dorsal scales are in 13 rows throughout the body. Three species occur in Thailand.

KEY TO THE THAI *CALAMARIA*

1	**a**	Five or six supralabials	**2**
	b	Four supralabials	*C. pavimentata*
2	**a**	1st infralabials in contact with each other behind the mental; supra-ocular less than half as broad as frontal; head is pink	*C. schlegeli*
	b	First infralabials not in contact; supra-ocular at least half as broad as frontal	*C. lumbricoidea*

Key prepared by Chan-ard, Parr and Nabhitabhata.

210 Variable Reed Snake *Calamaria lumbricoidea*

Body Length: Up to 38.8 cm.

A small, cylindrical snake. Dorsum is black or dark brown. The top of the head is brownish-lavender. A wide ivory yellow stripe is found on the flanks from the eye to the vent. The venter is ivory yellow with numerous blackish and slate bars. The infralabials, chin, and throat are ivory. The juvenile is black with yellow stripes, with a red or pink head. **Behaviour:** Nocturnal and fossorial. Lives in surface debris and leaf litter. **Habitat:** Found in Malayan mixed dipterocarp forest. **Range:** Found in the extreme south of peninsular Thailand. Known in Pattani and Narathiwat provinces. **Regional Range:** Thailand, Peninsular and Eastern Malaysia, Singapore, Brunei, Indonesia, and the Philippines.

210

211 Collared Reed Snake *Calamaria pavimentata*

Body Length: Up to 16.2 cm.

A diminutive snake. Reddish-brown to very-dark brown above with five rows of tiny, deep black spots. The neck has a black collar followed by a yellow ring. The head is dark, while the chin and throat are white. The venter is white or yellow. An indistinct median line is found under the pointed tail. The ventrals number 133-168 and the sub-caudals number 13-20.

Behaviour: Fossorial habit. **Habitat:** Found in sub-montane forest. **Range:** Present in northern and southeastern Thailand in mountainous areas. **Regional Range:** Myanmar, China, Vietnam, Laos, Cambodia, Thailand, Peninsular Malaysia, Singapore, and Japan (Ryukyu Islands).

211

212 Red-headed Reed Snake *Calamaria schlegeli*

Body Length: Up to 45.0 cm.

A large calamarian with very smooth scales. The head is orangy-red which strongly contrasts with the iridescent bluish-black dorsal body. The chin, throat, and venter are ivory-white. The sub-caudals have dim lavender transverse bars on each scale. The tail is pointed. **Behaviour:** Fossorial, living in surface debris. Feeds on worms, slugs, and insect larvae. Also consumes frogs. **Habitat:** Found in Malayan mixed dipterocarp forest. **Range:** Occurs in extreme south of peninsular Thailand. One record from Pattani Province. **Regional Range:** Thailand, Peninsular and Eastern Malaysia, Singapore, and Indonesia.

212

Reed Snakes Genus *Pseudorabdion*

This genus comprise terrestrial fossorial species living in leaf litter and forest debris. They feed on small lizards, frogs, worms, and insects. They are small, harmless snakes. The eyes are small, with small pupils. The sub-caudals are usually arranged in pairs. They are oviparous. They have 15 dorsal scale-rows. One species occurs in Thailand.

213 Long-headed Reed Snake *Pseudorabdion longiceps*

Body Length: Up to 20.8 cm.

A diminutive snake with varied body coloration. Dorsal coloration may be uniformly slaty grey, brownish red, black, or bluish grey above. The top of the head is darker than the sides. There is usually a white or yellowish collar around the neck. The ventrals are greyish-brown. The scales are in 15 rows throughout the body. **Behaviour:** Nocturnal. Mainly fossorial habit. **Habitat:** Found in Malayan mixed dipterocarp forest. **Range:** Restricted to Narathiwat and Yala provinces in extreme southernmost peninsular Thailand. **Regional Range:** Thailand, Peninsular and Eastern Malaysia, Singapore, Brunei, and Indonesia (Kalimantan and Sumatra).

213

Mountain Reed Snakes Genus *Macrocalamus*

Allied to *Calamaria* but scale reduction has proceeded less far, there being 15 scale-rows. There are seven or eight upper labials and a temporal shield separates the posterior ones from contact with the parietals. A loreal shield is usually present and the nostril is pierced between a large nasal and the first upper labial. Anal entire. One species occurs in Thailand.

214 Malayan Mountain Reed Snake *Macrocalamus lateralis*

Body Length: Up to 40 cm.

Mid-brown above with or without oblique yellow lines or more-irregular markings on the head, neck, and forepart of the body. Yellow or orange below, the ventrals tipped with dark brown; lowest row of scales often pale forming a light lateral streak. Ventrals 109-132, sub-caudals 18-28; scales in 15 rows. Eight upper labials, 4th and 5th touching the eye; a loreal present but no internasals. Head small and wedge-shaped, tail tapering and sharply pointed. **Behaviour:** Diurnal. Mainly fossorial habit, but also terrestrial.

Habitat: Recorded at 1,200 m in sub-montane peninsular monsoonal evergreen forest, in Khao Luang, Nakhon Si Thammarat Province and in lowland Malayan mixed dipterocarp forest in Narathiwat Province. **Range:** Restricted to southern peninsular Thailand. **Regional Range:** Thailand and Peninsular Malaysia.

214

Bronzebacks Genus *Dendrelaphis*

The bronzebacks are moderately long, extremely slender arboreal snakes which are active by day. They have elongate heads which are distinct from the body and large eyes with a round pupil. They feed on frogs and lizards. All are oviparous. Seven species occur in Thailand.

KEY TO THE THAI *DENDRELAPHIS*

1 a	13 dorsal scale-rows at mid-body	*D. caudolineatus*
b	15 dorsal scale-rows at mid-body	**2**
2 a	A single elongate supralabial forming lower border of orbit; vertebrals feebly enlarged, narrower than scales of outer row; seven (or more rarely eight) supraoculars	**3**
b	More than a single labial entering orbit	**4**
3 a	A light ventrolateral stripe present	*D. pictus*
b	No light ventrolateral stripe	**4**
4 a	Temporal stripe covers only the lower part of the temporal region and ends at the rear of the jaw; neck red when inflated; eye large; two supralabials touch the eye; first supralabial touches two infralabials	*D. kopsteini*
b	Temporal stripe covers only the whole temporal region	**5**
5 a	Ventrals less than 170; black oblique bars laterally on the body	*D. striatus*
b	Ventrals more than 170; flanks without oblique bars	**6**

6 a	Supralabials and adjoining neck region yellow; flanks olive- bronze	*D. ngansaoensis*
b	Neck not yellow; flanks distinctly bluish	*D. cyanochloris*

Key modified from Taylor (1965); Vogel & Rooijen (2007).

215 Striped Bronzeback *Dendrelaphis caudolineatus*

Body Length: Up to 118.2 cm.

A very slender, moderately long, snake. The top of the head is uniformly olive-brown, with greenish-yellow supralabials. The chin and throat are ivory-yellow. The body is olive-brown anteriorly; the body coloration develops a whitish stripe bordered by two narrower jet-black stripes. The tail tapers gradually to a fine point. The venter is greyish-green. The sub-caudals have a black median line. **Behaviour:** Mainly arboreal habit. Often seen in bushes or trees in well-watered forests. Feeds on frogs, rats, and other small rodents. **Habitat:** Inhabits peninsular monsoonal evergreen forest and Malayan mixed dipterocarp forest up to 1,100 m. **Range:** Present in peninsular Thailand northward to the Kra Isthmus. **Regional Range:** Myanmar, Thailand, Peninsular and Eastern Malaysia, Singapore, Brunei, Indonesia, and the Philippines.

215

216 Mountain Bronzeback *Dendrelaphis subocularis*

Body Length: Up to 62.9 cm.

A very slender, small bronzeback. The head and the anterior part of the neck are bronze-brown; this coloration extends to the tip of the tail. A cream or pale-brown stripe extends along both sides bordering the ventrals. The ventrals are white. The chin is creamish with lavender wash. **Behaviour:**Arboreal. Diurnal. They feed on frogs and lizards. **Habitat:** Found in a wide range of forest types, up to 450 m. **Range:** Present throughout

Thailand, though status in northeastern Thailand is unconfirmed. Absent from the forests of the peninsula south of Chumphon Province. **Regional Range:** Myanmar, China, Vietnam, Laos, Cambodia, and Thailand.

216

217 Elegant Bronzeback *Dendrelaphis kopsteini*

Body Length: Up to 142.0 cm.

Extremely slender and very long. The neck and sometimes the head are reddish-brown. Bronze, brown, or slaty-grey above. Small green or blue spots may be present on the sides. A black stripe on the sides of the head passes through the very large eyes, then widens considerably onto the nape and converges. Two black stripes may be present along the sides of the body posteriorly. The ventral surface is light green. The scale-rows at mid-body

217

number 15. **Behaviour:** Arboreal and diurnal. Feeds on frogs and lizards. **Habitat:** Inhabits peninsular monsoonal evergreen forest and Malayan mixed dipterocarp forest. **Range:** Restricted to provinces of southern peninsular Thailand. **Regional Range:** Thailand, Peninsular Malaysia, Singapore, and Indonesia (Sumatra).

218 Wall's Bronzeback *Dendrelaphis cyanochloris*

Body Length: Up to 133.0 cm.

A very long, slender bronzeback. Slaty-grey or bluish-grey above with a characteristic net-like pattern formed by the black edges of the smooth body scales. Usually lacks a black flank stripe. A broad black temporal stripe extends onto the body where it breaks into spots. The lips and lower jaw are yellowish. The ventrals and outer dorsal scale-rows are pale greenish or yellowish. The ventrals number 186-211. **Behaviour:** Diurnal. Largely arboreal. Diet consists of birds, lizards, and frogs. **Habitat:** Found in peninsular monsoonal evergreen forest and Malayan mixed dipterocarp forest. **Range:** Recorded in peninsular Thailand. **Regional Range:** India (including Andamans and Nicobar Islands), Bangladesh, Myanmar, Thailand, and Peninsular Malaysia.

218

219 Common Bronzeback *Dendrelaphis pictus*

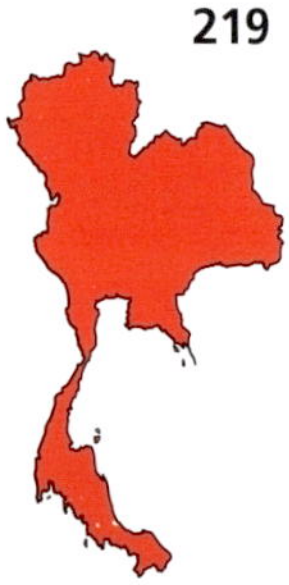

Body Length: Up to 125.0 cm.

Slender-bodied snake with a bronze-brown dorsum and top of head. The head has a black band passing through the large eye onto the neck and running down the length of the body as a narrow black stripe. Below this black stripe, a creamy-yellow stripe runs down each flank to the vent, bordered below by a black stripe of equal width. It can dilate its body when excited, showing pale-blue interstitial skin on neck. The upper labials are white. The venter is pearly-white. **Behaviour:** Almost entirely arboreal and extremely agile. Diurnal, feeding on frogs and lizards. **Habitat:** Found in a wide range of forest types. **Range:** Widespread throughout the country. **Regional Range:** India, Nepal, Bangladesh, Myanmar, China, Vietnam, Laos, Cambodia, Thailand, Peninsular and Eastern Malaysia, Singapore, Brunei, Indonesia, and the Philippines.

219

---

220 Vietnamese Bronzeback ***Dendrelaphis ngansonensis***

Body Length: Up to 143.0 cm.

Dorsum is slate-grey, bronze-brown or brownish-olive. The forehead is brown with a black band passing through the eye onto the neck. The upper labials are white. The venter is white. **Behaviour:** Almost entirely arboreal and extremely agile. Diurnal. Feeds on frogs and lizards. **Habitat:** Found

220

in southeastern monsoonal evergreen forest. **Range:** Recorded in Namtok Plieu National Park, Chanthaburi Province in southeastern Thailand. **Regional Range:** Laos, Vietnam, and Thailand.

221 Banded Bronzeback *Dendrelaphis striatus*

Body Length: Up to 100.0 cm.

The diagnostic dorsal patterning is bronze-coloured closest to the head, changing to a deep blue-and-white pattern in the middle of the body, which fades to rich brown tones near the tail. The entire dorsal surface is marked with a series of narrow, oblique black bars on each side of the body. The lateral black and white stripes found in most other Bronzeback species are absent. The head is bronze-coloured with a black stripe running from the snout through the eye and fading just behind the neck. **Behaviour:** Arboreal and diurnal. **Habitat:** Inhabits peninsular monsoonal evergreen forest and Malayan mixed dipterocarp forest. **Range:** Found in peninsular Thailand. **Regional Range:** Thailand, Peninsular and Eastern Malaysia, and Indonesia.

221

Tree Snakes Genus *Chrysopelea*

This group is known as the flying snakes, comprising some of the most arboreal snakes. The ventrals are keeled laterally and notched. The head is weakly flattened, distinct from the neck. They are mildly venomous. In Thailand 3 species are represented.

KEY TO THE THAI *CHRYSOPELEA*

1 a	Green and black body coloration	**2**
b	Brick-red upper body with black-edged white bars	*C. pelias*
2 a	Body is green-coloured, with each scale with a median black line; head banded in black and yellow; body often banded with black	*C. ornata*
b	Black above, with each scale with a central yellow spot; often has a median dorsal series of red or yellow flower-shaped spots	*C. paradisi*

Key prepared by Chan-ard, Parr and Nabhitabhata.

222 Paradise Tree Snake *Chrysopelea paradisi*

Body Length: Up to 77.4 cm.

Superficially similar in appearance to the Golden Tree Snake. The body is a kaleidoscope of colour, the result of each scale on the body being black with a green spot. The head is black crossed by five greenish-yellow bands; the interspaces are covered with a series of greenish-yellow spots. The venter is greenish-yellow. The ventrals are edged in black. **Behaviour:** Arboreal. It is highly agile, and able to glide considerable distances by inverting its ventral surface. Feeds on geckos. **Habitat:** Occurs in peninsular monsoonal evergreen forest and Malayan mixed dipterocarp forest. **Range:** Found in peninsular Thailand. **Regional Range:** India (Andamans), Myanmar, Thailand, Peninsular and Eastern Malaysia, Singapore, Brunei, Indonesia, and the Philippines.

222

223 Golden Tree Snake *Chrysopelea ornate*

Body Length: Up to 106.8 cm.

An attractive medium-sized snake which is similar in appearance to the Paradise Tree Snake. The body is dark green, with each greenish-yellow scale having a median black line and a black edge. The black head has four narrow, greenish-yellow, transverse bands; the interspaces are covered with a series of greenish-yellow spots. The chin and supralabials are ivory coloured. The ventral and sub-ventrals are greenish-yellow with small black dots. The eyes are large with round pupils. **Behaviour:** Arboreal and diurnal. Diet consists of lizards, particularly agamids and geckos. **Habitat:** Found in all forest types. **Range:** Found throughout the country. **Regional Range:** India, Myanmar, China, Vietnam, Laos, Cambodia, Thailand, and Peninsular Malaysia.

223

224 Twin-barred Tree Snake *Chrysopelea pelias*

Body Length: Up to 74.0 cm.

This Tree Snake is identified by its brick-red upper body and black-edged white bars. The black or brown head has three contrasting red, orange, or yellowish-orange bars transversing it. The first and narrowest of these three bars crosses just in front of the eyes; the second passes just behind the eyes, with the third and widest crossing the base of the head. The flanks are light-brown speckled with white. The ventral surface is ivory white. **Behaviour:**Arboreal and diurnal. Able to glide considerable distances by

224

inverting its ventral surface and launching itself from the tree tops. Feeds largely on lizards. **Habitat:** Found in peninsular monsoonal evergreen forest and Malayan mixed dipterocarp forest. **Range:** Restricted to peninsular Thailand. Only recorded in Khao Luang, Nakhon Si Thammarat Province and Ton Nga Chang, Songkhla Province. **Regional Range:** India (Andamans), Myanmar, Thailand, Peninsular and Eastern Malaysia, Singapore, Brunei, and Indonesia.

Wolf Snakes Genus *Lycodon*

This genus comprises small- to medium-sized slender snakes. The head is depressed. The eyes are relatively small, with vertical-oval pupils. The dorsals may be keeled or not, and are in 17, 19, or 21 rows. The sub-caudals are arranged in two rows. They are crepuscular or nocturnal, and all are ground-dwelling. They feed mainly on snakes, lizards, and frogs, as well as insects. Seven species occur in Thailand.

KEY TO THE THAI *LYCODON*

1	**a**	Either the loreal or the preocular absent	**2**
	b	Loreal and preocular scales both present	**3**
2	**a**	Preocular absent, both prefrontal and loreal scales in contact with eye, anal divided	*L. subcinctus*
	b	Loreal absent, prefrontal in contact with the 2nd and 3rd supralabials, anal entire; restricted to southern Thailand	*L. effraenis*
3	**a**	Anal divided	**4**
	b	Anal single	**5**
4	**a**	Anterior chin shields not more than 1.5 times longer than posterior ones; loreal in extensive contact with internasal, when adult no crossbands on body	*L. capucinus*
	b	Anterior chin shields 2-3 times longer than posterior ones; loreal not or in short contact with internasal (very rarely an extensive contact); yellow crossbands on forepart of body in adults	*L. laoensis*
5	**a**	Loreal entering the orbit, more than 20 light bands on body	**6**
	b	Loreal excluded from orbit or punctually in contact, 12 to 13 light bands on body; Eastern Thailand	*L. cardamomensis*
6	**a**	One anterior temporal scale; Southern Thailand	*L. butleri*
	b	Two anterior temporal scales	*L. fasciatus*

Key modified from Pauwels et al. (2005).

225 Common Wolf Snake *Lycodon capucinus*

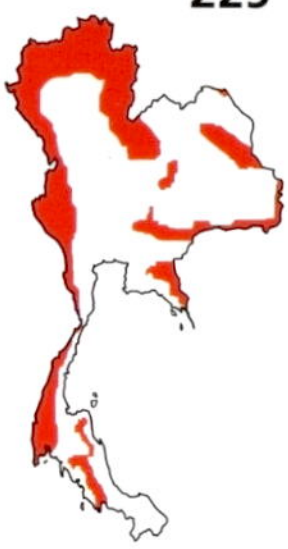

Body Length: Up to 76.0 cm.

Dark purplish-black head and snout are both fairly flattened. Has a white or yellow nuchal band. The body is lilac to lavender-brown with a reticulum of white. Tail is uniformly brown. Dirty-white ventrals are strongly angulate. **Behaviour:** Nocturnal and terrestrial. **Habitat:** Found in lower and upper montane evergreen forest. **Range:** Found in mountainous areas throughout continental and peninsular Thailand. **Regional Range:** Maldives, India (Andamans and Nicobar Islands), Myanmar, Vietnam, Laos, Cambodia, Thailand, Peninsular and Eastern Malaysia, Singapore, Indonesia, and Philippines.

225

226 Malayan Banded Wolf Snake *Lycodon subcinctus*

Body Length: Up to 92.3 cm.

This species changes colour with age. Young have 10-13 purplish-black bands separated by white crossbars. These white bars gradually fade in the posterior part of the body and are absent in the adults. The chin and throat are greyish-white. Most ventrals are greyish, with the sub-caudals being slightly darker grey. Scale-rows in 17 rows. Ventrals 197-230. **Behaviour:** Nocturnal and terrestrial. Feeds on lizards. **Habitat:** Found in dry evergreen forest, monsoonal evergreen forest, Malayan mixed dipterocarp forest, and peat swamp forest. **Range:** Present throughout Thailand. **Regional Range:** China, Vietnam, Laos, Cambodia, Thailand, Peninsular and Eastern Malaysia, Brunei, Indonesia, and the Philippines.

226

Juvenile

227 Yellow-barred Wolf Snake *Lycodon laoensis*
Body Length: Up to 47.5 cm.

A small snake. Body is deep bluish-grey, with narrow 13-26 bright-yellow or white crossbars. The spacing between the bars decreases along the body and tail. The head is moderately distinct from neck and slightly flattened; the top of the head is ultramarine blue. **Behaviour:** Nocturnal and terrestrial. Diet includes lizards and frogs. **Habitat:** Found in evergreen forests from

227

mid-altitude to the highest mountains. **Range:** Found throughout Thailand. **Regional Range:** India, China, Vietnam, Laos, Cambodia, Thailand, and Peninsular Malaysia.

---

228 Butler's Wolf Snake *Lycodon butleri*

Body Length: Up to 83.0 cm.

Blackish brown above with 40-50 irregular and more-or-less darkly suffused white crossbars, which are distributed throughout the length of the snake—widely spaced anteriorly, closely posteriorly. Whitish below with dark brown crossbars corresponding to, but narrower than, the dark interspaces of the dorsal pattern; these ventral crossbars may fail anteriorly or posteriorly or both and be replaced by irregular brown markings, but they are always distinct at mid-body. Scales in 17 rows at mid-body. **Behaviour:** Terrestrial and nocturnal habits. **Habitat:** Found in peninsular monsoonal evergreen forest and Malayan mixed dipterocarp forest in lowland areas. **Range:** Present in Peninsular Thailand. **Regional Range:** Thailand and Peninsular Malaysia.

---

229 Brown Wolf Snake *Lycodon effraenis*

Body Length: Up to 100.0 cm.

A medium-sized snake, with a reddish-brown or brown dorsum. The venter is an unpatterned brown. The flattened head has cream or yellow streaks. The juvenile has three broad cream or yellowish rings. **Behaviour:** Nocturnal and terrestrial. Diet consists of lizards and small snakes. **Habitat:** Found in lowland Malayan mixed dipterocarp forest. **Range:** Found in extreme south of Thailand, recorded from Yala and Narathiwat provinces. **Regional Range:** Thailand, Peninsular and Eastern Malaysia, Singapore, and Indonesia (Kalimantan and Sumatra).

229

230 Indian Banded Wolf Snake *Lycodon fasciatus*

Body Length: Up to 89.5 cm.

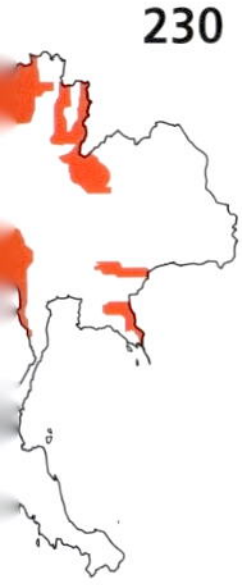

A medium-sized snake. Body is glossy black, with 22-48 narrow bright yellow or white crossbars. The spacing between the bars decreases along the body and tail. The flattened black head is moderately distinct from neck. **Behaviour:** Nocturnal and arboreal. **Habitat:** Found in mixed deciduous and dry evergreen forest. **Range:** Found throughout continental Thailand, southward to Petchaburi Province. **Regional Range:** India, Bangladesh, Myanmar, China, Vietnam, Laos, and Thailand.

230

231 Cardamom Wolf Snake *Lycodon cardamomensis*

Body Length: Up to 89.6 cm.

A small snake. Dorsal surface of body and tail black, with 13 pinkish-orange bands across the body (excluding the nuchal collar) and 6 across the tail. Median part of these bands speckled with dark brown. Belly whitish with central black specks increasing in density posteriorly, giving a black posterior ventral colour, but less at the places where the dorsal rings join the belly. Head distinct from neck, markedly flattened. Anal single. Dorsal scales in 17 rows at mid-body. **Behaviour:** Nocturnal and terrestrial. **Habitat:** Found in southeastern monsoonal evergreen forest, in lowland and hilly areas. **Range:** Currently recorded only in Chanthaburi province, southeastern Thailand. **Regional Range:** Cambodia and Thailand.

231

Wolf Snakes Genus *Dinodon*

This genus includes small to medium-sized snakes, with heads barely distinct from the neck. The eye is moderate with a vertically elliptical pupil. The body is elongate with the dorsal scales in 15 or 17 rows. The tail is long and the sub-caudals are paired. A single species is represented in Thailand.

232 Hill Wolf Snake *Dinodon septentrionalis*

Body Length: Up to 118.0 cm.

Head is fairly broad and flattened, with the front portion black and the rest of the head sometimes white. Diagnostic body coloration, having a purplish-tinged ground colour with 25-35 narrow white or yellowish bars, which become white on the sides. The venter is whitish, sometimes spotted with black. The tail is black below. It has 207-212 ventrals with lateral keels, and 81-92 caudals. **Behaviour:** Nocturnal and terrestrial, feeding on lizards and frogs. **Habitat:** Found in dry evergreen forest. **Range:** Found in Chiang Mai and Nan provinces, northern Thailand. **Regional Range:** India, Myanmar, China, Vietnam, Laos, and Thailand.

232

Wolf Snakes Genus *Lepturophis*

This genus comprises extraordinarily slim cylindrical snakes. The depressed head is very distinct from the body. The eye is moderately large with a vertical elliptical pupil. The dorsals have distinct keels in 17 rows. The tail is extremely long and thin. The sub-caudals are in two rows. These snakes are nocturnal and crepuscular. The genus is monotypic, with a single species occurring in Thailand.

233 Slender Wolf Snake *Lepturophis albofuscus*

Body Length: Up to 150 cm.

Dark-grey to brown above. The venter is unpatterned yellowish-white. Juveniles have 30-40 narrow white or yellow bands on the dorsum. Ventral scales 225-259, keeled and notched, sub-caudals 113-206, anal divided; scales in 17 rows, keeled, the keels finely serrated. Eight upper labials, 3rd to 5th touching the eye; a small preocular in contact with the frontal below which the loreal extends back to the eye. **Behaviour:** Nocturnal. Both terrestrial and arboreal, but most usually encountered on the ground. Feeds on skinks and frogs. **Habitat:** Found in lower and upper montane evergreen forest. Often found along stream beds. **Range:** Found in peninsular Thailand. **Regional Range:** Thailand, Peninsular and Eastern Malaysia, and Indonesia.

233

Red Snakes Genus *Stegonotus*

This genus comprises small to medium-sized snakes. The body is long, round or slightly compressed. The head is distinct from the neck. The dorsals usually have apical pits. They are nocturnal terrestrial snakes, being semi-fossorial. One species is represented in Thailand.

234 Borneo Red Snake *Stegonotus borneensis*

Body Length: Up to 71.2 cm.

Head is dark olive to brown with a pair of blackish markings surrounded by tiny yellow dots. Body is brownish-grey with an indistinct median row of tiny black dots and lighter dots running along the vertebral ridge. All ventral scales have black spots on outer edges; anterior ventrals also have a pair of black dots in the centre. **Behaviour:** Nocturnal and terrestrial habit, partly semi-fossorial. Feeds on snakes, lizards, reptile eggs, and small frogs. **Habitat:** Occurs in peninsular monsonal evergreen forest. **Range:** Recorded in Khao Luang National Park, Nakhon Si Thammarat Province in peninsular Thailand. **Regional Range:** Thailand and Eastern Malaysia (Sarawak).

234

Bridle Snakes Genus *Dryocalamus*

The bridle snakes are extremely slender snakes. The head is sub-ovate, and compressed. The eyes are large or very large with vertical pupils. They are arboreal in habit and usually nocturnal. They feed upon cold-blooded prey. Two species are represented in Thailand.

KEY TO THE THAI *DRYOCALAMUS*

1 a Dorsal scale-rows at mid-body 13; black above, banded with white or pale green — *D. davisonii*

b Dorsal scale-rows at mid-body 15; light-brown above with a more or less regular series of large transverse dark-brown spots, usually unbroken over the back but not reaching to the ventrals, so that the light colour is continuous along the sides; in sub-species thungsongensis, body dark brown, with a mediodorsal light stripe beginning above 7th ventral, uninterrupted, limited to vertebral row — *D. subannulatus*

Key prepared by Chan-ard, Parr and Nabhitabhata.

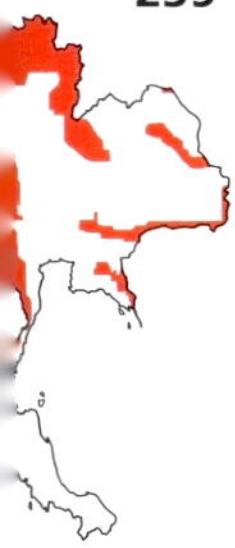

235 Common Bridle Snake *Dryocalamus davisonii*

Body Length: Up to 92.0 cm.

A long, extremely slender snake. It is blackish above, with 33-40 white bands which extend halfway down the sides of the body; the distance between these bands decreases posteriorly. The tail has 27-32 black bands. The ventral surface is white to greenish. The head is white with a light brown mark covering the frontal region to snout, and a median posterior mark reaching to the first dorsal black blotch. The eyes are large and the pupils round. The upper lip, chin, and throat are white. The tail is heavily speckled with brown. **Behaviour:** Nocturnal and terrestrial. It hunts lizards. **Habitat:** Inhabits lowland forested areas up to 300 m. **Range:** Recorded from continental and peninsular Thailand, southward to at least Krabi Province. **Regional Range:** Myanmar, Vietnam, Laos, Cambodia, and Thailand.

235

236 Malayan Bridle Snake *Dryocalamus subannulatus*

Body Length: Up to 50.0 cm.

A very slender, small snake. Occurs in two colour forms. In both forms the top of the head is brown with a yellow mask around the eye; both forms also have white labials, chin, throat, and ventrals. However, the dorsal coloration varies. In one form the body patterning is chocolate-brown with a golden-yellow stripe extending down the vertebral ridge. A second golden-yellow stripe extends down the side of the body. In the second form, the patterning consists of chocolate-brown bands, which extend halfway down the sides; the rest of the body is yellow. The spacing between these crossbands decreases posteriorly. On the lower sides of the body, small brown blotches are present on the anterior two-thirds of the body. The scales are smooth and are in 15 rows at mid-body. **Behaviour:** Nocturnal and arboreal. Feeds on lizards and frogs. **Habitat:** Inhabits peninsular monsoonal evergreen forest and Malayan mixed dipterocarp forest. **Range:** Recorded from southern peninsular Thailand. **Regional Range:** Thailand, Peninsular Malaysia, Singapore, Indonesia (Sumatra), and the Philippines (Palawan).

236

Mountain Snakes Genus *Plagiopholis*

The mountain snakes are small, slender snakes. The head is scarcely distinct from the body. They are terrestrial in habit and usually nocturnal. They feed on earthworms. They are oviparous. Two species are present in Thailand.

KEY TO THE THAI *PLAGIOPHOLIS*

1 a	Scales scarcely oblique; five supralabials, 3rd touching the eye	*P. blakewayi*
b	Scales distinctly oblique; six supralabials, 3rd and 4th touching the eye	*P. nuchalis*

Key modified from Smith (1935).

237 Burmese Mountain Snake *Plagiopholis nuchalis*

Body Length: Up to 47.0 cm.

A small cylindrical snake similar in appearance to a kukri snake. Blackish-brown or reddish-brown above, with many of the scales edged with black. It has a dorso-lateral series of rounded black spots connected by light-brown crossbars. The dark brown head is indistinct. It has a broad black chevron with its apex pointing toward the head. The eyes have vertically elliptical pupils. There are two dark marks below the eye. The chin and throat are yellowish. Yellowish or pinkish below, speckled with black. **Behaviour:** Terrestrial. It feeds almost exclusively on earthworms. **Habitat:**

Found in sub-montane forest and evergreen forest from 655-1,620 m. **Range:** Inhabits northern, western, and northeastern (west) Thailand. Recorded in Mae Hong Song, Chiang Mai, Lampang, Loei, and Kanchanaburi provinces. **Regional Range:** India, Myanmar, China, Vietnam, and Thailand.

237

238 Blackeway's Mountain Snake *Plagiopholis blackewayi*

Body Length: Up to 47.0 cm.

Dorsum is greyish-brown. It has a narrow blackish band on the nape. The body and tail are marked with a longitudinal row of small black-brown blotches. The venter is greyish-cream. Clearly distinguishable from Burmese Mountain Snake by the absence of a loreal shield, No or one posterior temporals (two in Burmese Mountain Snake), and five supralabials (six in Burmese Mountain Snake). The 3rd supralabial touches the eye and the 4th is the largest one; by comparison in the Burmese Mountain Snake the 3rd and 4th supralabials are in contact with the eye and the 5th is the largest. **Behaviour:** Fossorial, feeding on earthworms. **Habitat:** Found in sub-montane forest and uppermontane forest from 1,520 to 2,200 m. **Range:** Recorded in Doi Pa Hom Pok, Chiang Mai Province in northern Thailand. **Regional Range:** Myanmar, China, and Thailand.

Blackheads Genus *Sibynophis*

The genus are ground-dwelling, diurnal snakes. They are small and slender with the head moderately distinct from the body. The eyes are large and the pupils are round. There is a single loreal shield. The teeth are numerous and evenly sized. The tail is long and slender. All are oviparous. They have 17 rows of dorsal scales at mid-body. Three species are found in Thailand.

KEY TO THE THAI *SIBYNOPHIS*

1 a Black triangular nuchal spot, bordered laterally by a cream or yellow stripe continuous with cream lines on labials; 10th supralabial much the largest *S. triangularis*

b Nuchal dark band not triangular; 10th supralabial not the largest **2**

2 a Ten supralabials, 8th much the largest; two post-oculars, only upper touches parietal *S. collaris*

b Brown or blackish above with two lighter yellowish or bluish-grey longitudinal streaks, each of which may bear a series of black spots; nine supralabials; two post-oculars, with both touching parietal *S. melanocephalus*

Key modified from Smith (1935); Taylor (1965).

239 Malayan Blackhead *Sibynophis melanocephalus*

Body Length: Up to 24.6 cm.

A medium-sized snake. Head and occipital region is reddish with faint light flecks. It has nine supralabials. The cylindrical body has a median brown or blackish stripe on the vertebral ridge; on either side of this is a lighter yellowish-brown or bluish-grey longitudinal stripe; each of these may in turn bear a series of black bars. This pattern fades toward the tail. Throat and venter is yellowish white. Scales in 17 rows. **Behaviour:** Diurnal and terrestrial habit. Occurs near streams, feeding on lizards, frogs, tadpoles, and large insects. **Habitat:** Found in Malayan mixed dipterocarp forest. **Range:** Restricted to southernmost provinces, preferring lowlands. Recorded in Pattani and Narathiwat provinces. **Regional Range:** Laos, Thailand, Peninsular and Eastern Malaysia, Singapore, Indonesia, and Philippines (Tawitawi).

239

240 Triangle Blackhead *Sibynophis triangularis*

Body Length: Up to 32.7 cm.

Head is variegated olive, with a black triangular nuchal spot, bordered laterally by a diagnostic bright-cream or yellow stripe. The cylindrical body is generally greyish-brown with a light-brown line running from the neck backward (on 4th and part of 5th scale-row). On the vertebral ridge a row of small black dots extends a short distance; farther back, a discontinuous series of white spots continue backward, fade, and reappear on the tail. Ventral surface is whitish with minute black dots on outer edge of each ventral and sub-caudal. **Behaviour:** Diurnal and terrestrial. **Habitat:** Found in southeastern monsoonal evergreen forest. **Range:** Known from southeastern Thailand. **Regional Range:** Endemic to Thailand.

240

241 Collared Blackhead *Sibynophis collaris*

Body Length: Up to 71.2 cm.

Medium-sized. Head is dark-olive to brown with a pair of blackish markings surrounded by tiny yellow dots. Body is brownish grey with an indistinct median row of tiny black dots and lighter dots running along the vertebral ridge. All ventral scales have black spots on outer edges; anterior ventrals also have a pair of black dots in the centre. **Behaviour:** Diurnal and terrestrial. Diet consists of snakes, skinks, frogs, and insects. **Habitat:** Found in dry evergreen forest. **Range:** Scattered distribution. Currently recorded in northern, eastern central, and southwestern Thailand. Recent records from Chiang Mai, Nakhon Ratchasima, and Kanchanaburi provinces. **Regional Range:** India, Nepal, Myanmar, China, Vietnam, Laos, Thailand, and Peninsular Malaysia.

241

Cat Snakes Genus *Boiga*

The Cat Snakes are long and slender, with very noticeably, vertically compressed, bodies. They have large rounded heads, large eyes, and vertical slit-like pupils. They are nocturnal and primarily arboreal. Their prey includes small mammals, birds, eggs, and other reptiles including snakes. They are rear-fanged snakes, not considered dangerous to humans. They are oviparous. Thailand is represented by 9 species.

KEY TO THE THAI *BOIGA*

1 a	Scale-rows at mid-body number 19	**2**
b	Scale-rows at mid-body number 21 or more	**3**
2 a	A large species reaching 2 m in length; A pair of black stripes on venter; eye as long as snout; sub-caudals 114-169	*B. drapiezii*
b	Small species reaching 1 m in length; no stripes; venter whitish, marbled or spotted with brown, with a series of brown spots along each side; sub-caudals number 80-106	*B. multomaculata*
3 a	Scale-rows number 21	**4**
b	Scale-rows number 23-25	**8**
4 a	Green above, with or without light transverse bars	**5**
b	Adults not green	**6**
5 a	A large species, reaching 1.75 m; uniform green, with black interstitial skin; sub-caudals 124-138; young greenish or greyish with indefinite darker pattern	*B. cyanea*
b	Body length reaching 1.6 m, bright yellow-green with light bands, interspaced with black blotches, black interstitial skin; sub-caudals 116-121	*B. saengsomi*

6	**a**	Body deep black with yellow vertical bars on sides of body and tail; snout longer than eye; sub-caudals number 89-110	*B. dendrophila*
	b	Body not black with yellow vertical bars	**7**
7	**a**	Head blackish; body reddish or pale greyish-brown; upper lip and venter yellowish-white	*B. nigriceps*
	b	Large species, reaching 2 m; body light brown or reddish-brown, mottled or speckled with brown or black, and with blackish crossbands interrupted on the median line; head shields have discrete black spots.	*B. jaspidea*
8	**a**	Large snake; uniform fawn with black chevron marks enclosing lighter marks on mid-dorsal line; a series of whitish marks on outer scale-rows; head nearly uniform fawn with a well-defined dark black stripe behind eye brown	*B. cynodon*
	b	Body length reaches 1.6 m; light brown with 94 black or dark-brown bands about two costal wide, interspaced with lighter bands	*B. siamensis*

Key modified from Taylor (1965).

242 White-spotted Cat Snake *Boiga drapiezii*

Body Length: Up to 210.0 cm.

A very large, slender snake reaching more than 2 m in length. Like all cat snakes, its body is vertically compressed. Dorsum varies in coloration, being either predominantly reddish-brown or olive-green. The vertebral region has paired pink spots anteriorly. There are cream spots on the flanks. The supralabials are almost white. The chin and throat are white. It has a pair of black stripes on the venter. The sub-caudals number 114-169. **Behaviour:** Arboreal and nocturnal. Diet consists of birds, their eggs, lizards, frogs, and large insects. **Habitat:** Found in peninsular monsoonal evergreen forests up to 1,100 m. **Range:** Peninsular Thailand. **Regional Range:** Thailand, Peninsular and Eastern Malaysia, Singapore, Brunei, Indonesia, and the Philippines (Tawitawi).

242

243 Many-spotted Cat Snake *Boiga multomaculata*

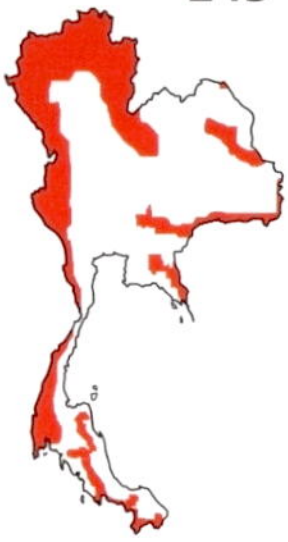

Body Length: Up to 187.0 cm.

One of the larger cat snakes, reaching almost 2 m long. The head is very distinct from the neck and is grey with a brown stripe beginning from the snout to the back of the head. A brown stripe also passes from the nostrils through the eye. The slender, vertically compressed body is brownish-grey above with more-or-less paired brown spots running the length of the body, each spot enclosed within a narrow creamy-white border. The sides have irregular spots or blotches usually alternating with the dorsal spots. The venter is marbled whitish, with small brown spots. **Behaviour:** Arboreal and nocturnal. Found from the lowlands up to 1,200 m. **Habitat:** Found in a wide range of forest types. **Range:** Found throughout the country in forested areas. **Regional Range:** India, Bangladesh, Myanmar, China, Vietnam, Laos, Cambodia, Thailand, Peninsular and Eastern Malaysia, Singapore, and Indonesia.

243

244 Green Cat Snake *Boiga cyanea*

Body Length: Up to 186.0 cm.

A large, slender snake. The vertically compressed body is uniformly olive-green above, with a bluish tinge. The head is also olive-green above. The eyes are large with vertical pupils. The chin and throat are bluish-white. The venter is greenish-white. The young are brownish-yellow. The ventrals number 237-257 and the sub-caudals number 124-158. **Behaviour:** Arboreal and nocturnal. Occasionally found on the forest floor. Diet includes small

244

mammals, birds and their eggs, other snakes, lizards, and frogs. **Habitat:** Found in a wide range of forest types from lowlands up to 2,100 m. **Range:** Found in all evergreen forests throughout the country. **Regional Range:** India, Nepal, Bangladesh, Myanmar, China, Vietnam, Laos, Cambodia, Thailand, and Peninsular Malaysia.

---

245 Banded Green Cat Snake *Boiga saengsomi*

Body Length: Up to 162.0 cm.

Body is predominantly pale green above, with black interstitial skin with paler greenish-yellow bands. The head is bright green above, the eight supralabials yellow with the black interstitial skin between each scale. The chin and throat are white. The venter is yellowish-white. The dorsal scales number 21 at mid-body. **Behaviour:** Nocturnal and arboreal. Found close to waterways. Diet consists of small mammals and birds. **Habitat:** Found in lowlands in peninsular monsoonal evergreen forest. Recorded in Krabi Province. **Range:** Found in peninsular Thailand. **Regional Range:** Endemic to Thailand.

245

---

246 Mangrove Cat Snake *Boiga dendrophila*

Body Length: Up to 250.0 cm.

A large, vertically compressed snake. It has a deep-black body and tail with narrow, lateral yellow bars extending to the venter but not crossing it. The head, which is very distinct from the body, is black above with bright-yellow labials, edged with black. The chin and the front part of the throat are yellow. There are 21-23 scale-rows around the mid-body. **Behaviour:** Nocturnal. Arboreal, found in trees over water. When disturbed, they slide into the water. Diet consists of birds, nestlings and birds' eggs, other snakes, lizards, and amphibians. **Habitat:** Inhabits peninsular monsoonal evergreen forest and Malayan mixed dipterocarp forest. **Range:** Found in southern peninsular Thailand, northward to Ratchaburi Province. Recorded in Ratchaburi, Nakhon Si Thammarat, Phatthalung, Trang, Pattani, and Narathiwat provinces. **Regional Range:** Thailand, Peninsular Malaysia, Singapore, Indonesia, and the Philippines.

246

247 Red Cat Snake *Boiga nigriceps*

Body Length: Up to 175.0 cm.

A large, slender snake. The body is uniformly straw-brown or irregularly mottled with black spots on the lower sides. Sometimes there are two longitudinal rows of fine black dots. The head is dark reddish-brown above. The upper lip and lower parts are creamish-white. The venter is pink or dark yellow; it lacks ventral stripes. It is oviparous. Juveniles are very pale

247

brown, with even paler banding. The scale-rows number 21 at mid-body. **Behaviour:** Arboreal and nocturnal. Diet consists of small mammals, birds, other snakes, lizards, and frogs. **Habitat:** Found in peninsular monsoonal evergreen forest and Malayan mixed dipterocarp forest. **Range:** Found in peninsuar Thailand; recent records from Pattani and Narathiwat provinces. **Regional Range:** China, Thailand, Peninsular and Eastern Malaysia, and Indonesia.

248 Jasper Cat Snake *Boiga jaspidea*

Body Length: Up to 150.0 cm.

A moderately large, vertically compressed, snake, somewhat thinner than other members of the genus. The dorsal body pattern is brown to reddish-brown, with two rows of dark spots or bars on the sides. It has a greyish-red vertebral stripe, with dark-brown spots to the sides. The lower sides have a series of white or yellow spots, alternating with indistinct black patches. The eye is very large with a vertical pupil. The head is brown to reddish-brown above. The labials are white with black spots. The chin and underside of the head are white. The venter is bright yellow anteriorly, fading to white toward the tail. The ventrals number 243-267 and the sub-caudals 140-166. The dorsal scales are 21 rows. **Behaviour:** Arboreal and nocturnal. Found along river banks and in other well-watered, humid environs. Feeds on lizards, but also small mammals and birds and their eggs. **Habitat:** Found in peninsular monsoonal evergreen and Malayan mixed dipterocarp forest. Occurs from sea level to about 800 m. **Range:** Found in peninsular Thailand northward to the Isthmus of Kra. **Regional Range:** Thailand, Peninsular and Eastern Malaysia, Singapore, and Indonesia.

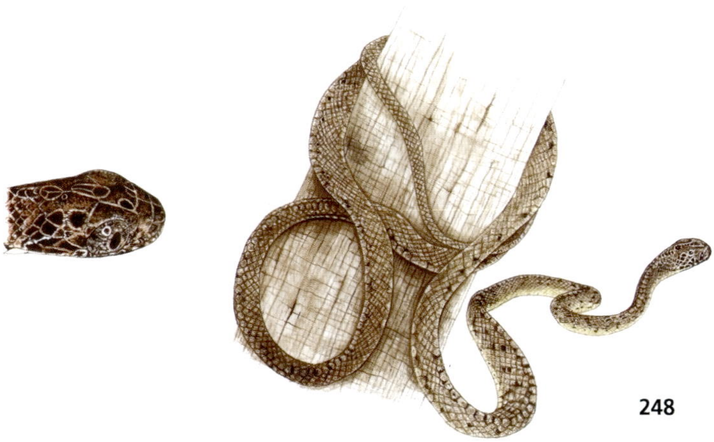

248

249 Grey Cat Snake *Boiga siamensis*

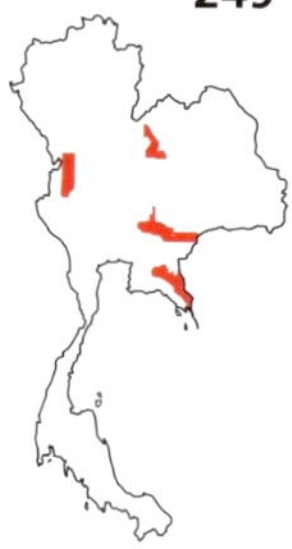

Body Length: Up to 170.0 cm.

The dorsal body pattern is greyish-brown with 87-98 black and brown narrow crossbars. Alternating dark and light spots extend along the sides. The head is distinct from the neck, being darker from the dorsal coloration of the body. The eye is large and has a vertical pupil. A dark mark extends from behind the eye to behind the last supralabial. Two dark lines extend from the base of the head to the first dark crossbar. The chin and throat are white. The ventrals are light-brown or grey, and number 247-270. The sub-caudals number 116-129. The anal plate is single. The dorsal scales are found in 23 rows at mid-body. **Behaviour:** Arboreal and nocturnal. Diet consists of small mammals and birds and their eggs. **Habitat:** Found in dry evergreen forest and mixed deciduous forest. **Range:** Found in western, northeastern (southern), southeastern, and central Thailand. **Regional Range:** India, Bangladesh, Myanmar, Vietnam, Laos, Cambodia, and Thailand.

249

250 Dog-toothed Cat Snake *Boiga cynodon*

Body Length: Up to 197.5 cm.

A large snake, close to 2 m in length. It has a uniformly fawn body with black chevron-shaped marks. The head is uniformly fawn with a well-defined black strip behind the eye. It has a series of whitish marks on the outer scale-rows. The ventrals are grey brown while the sub-caudals are black. Melanistic forms are common. **Behaviour:** Arboreal and nocturnal. Diet consists of small mammals, birds and their eggs, and lizards. **Habitat:** Lives

250

in peninsular monsoonal evergreen forest, Malayan mixed dipterocarp forest and peat swamp forest. **Range:** Found in peninsular Thailand south of the Isthmus of Kra. **Regional Range:** Thailand, Peninsular and Eastern Malaysia, Singapore, Indonesia, and the Philippines.

Mock Vipers Genus *Psammodynastes*

Members of this group have two enlarged, grooved maxillary teeth. In Thailand, the genus is represented by a single species, *Psammodynastes pulverulentus*.

251 Common Mock Viper *Psammodynastes pulverulentus*

Body Length: Up to 43.6 cm.

A small snake barely reaching 50 cm in length; occurs in two colour phases—either having a brown or reddish-brown dorsum. Both colour phases have numerous short, diagonally placed dots on the dorsum. The head is triangular, distinct from the neck and body, while the 3rd or 4th maxillary is enlarged and fang-like, leading to it often being mistaken for a viper. The chin, throat, and venter are white; it is greyish on the sub-caudal region. The large eye has a vertical pupil. **Behaviour:** Terrestrial. Both diurnal and nocturnal. Preys on small frogs and lizards. **Habitat:** Found in lower- and upper-montane forest above 700 m. **Range:** Present in upland areas throughout the country. **Regional Range:** India, Nepal, Bhutan, Bangladesh, Myanmar, China, Vietnam, Laos, Cambodia, Thailand, Peninsular and Eastern Malaysia, Singapore, Indonesia, and the Philippines.

251

Sand Snakes Genus *Psammophis*

This is a large genus. The head is distinct from the body. The pupils are round. The scales are smooth, more or less oblique in 17 rows. The tail is long. The sub-caudals are paired. Only a single species, *Psammophis indochinensis*, is represented in Thailand.

252 **Indochinese Sand Snake** *Psammophis indochinensis*

Body Length: Up to 110.0 cm.

A slender snake reaching over a metre in length. The cream-coloured head and body have three dark-brown, black-edged stripes which extend from the snout on the head along the entire body to the tail. The head is scarcely distinct from the neck. The eyes are large and rounded. The venter is dark-grey or white. **Behaviour:** Diurnal and largely terrestrial, but also occurs in understorey and undergrowth. Hunts rodents, lizards, and frogs. **Habitat:** Grasslands, open forests, and agricultural areast, from the plains up to 2,000 m. **Range:** Scattered distribution, with records from Doi Suthep National Park, Chiang Mai Province and Phu Khieo Wildlife Sanctuary, Chaiyaphum Province, as well as in the Central Plains region. **Regional Range:** Myanmar, Vietnam, Thailand, and Indonesia (Java and Bali).

252

Whip Snakes Genus *Ahaetulla*

These snakes are long, slender arboreal snakes. They have characteristically elongated, sharply pointed heads. A prominent ridge is present between the eyes and the snout. The eyes are transversely oval with a horizontal pupil. They are rear-fanged. Four species occur in Thailand.

KEY TO THE THAI AHAETULLA

1	**a**	Snout with a pointed dermal appendage, extending far beyond the lower jaw, and formed from rostral shield with, rarely, small separate scales at the base; anal divided; at least one supraocular divided horizontally	*A. nasuta*
	b	Snout lacking a pointed dermal appendage; supraoculars single	**2**

2 a Head with a pattern of black spots or dots; coloration brown or grey; more than 211 ventrals; more than 178 sub-caudals; anal usually single *A fasciolata*

b Top of head unmarked; coloration usually green, exceptionally other colours **3**

3 a More than 194 ventrals; more than 151 sub-caudals; in the green form, belly pale green with a yellow stripe along each side; anal only exceptionally single *A. prasina*

b Fewer than 195 ventrals; fewer than 156 sub-caudals; in the green form, belly usually white with a pair of longitudinal green lines and sometimes an additional green line along the middle; anal only exceptionally divided *A. mycterizans*

Key modified from Taylor (1965).

253 Oriental Whip Snake *Ahaetulla prasina*

Body Length: Up to 120.0 cm.

The largest and most abundant whip snake. It has a long head and pointed snout but lacks the appendage found in the Long-nosed Whip Snake. Body coloration is quite variable, usually being green of various shades dorsally, but may also be grey or orange and brown, speckled with black. The ventral is paler green. A yellowish or whitish line runs along the keels on each side of the ventrals. The head is not speckled. It has a divided anal shield. It has 194-235 ventrals and 151-187 sub-caudals; in the very similar Malayan Green Whip Snake the anal shield is single and the ventrals number 186-195 and the sub-caudals number 136-156. **Behaviour:** Arboreal and diurnal. It feeds on birds and lizards. **Habitat:** Occurs in the understorey and thick undergrowth. Found in a wide range of forest types up to 2,100 m. **Range:** Widespread in forested areas throughout the country. **Regional Range:** India, Bhutan, Bangladesh, Myanmar, China, Vietnam, Laos, Cambodia, Thailand, Peninsular and Eastern Malaysia, Singapore, Brunei, Indonesia, and the Philippines.

253

254 Malayan Green Whip Snake *Ahaetulla mycterizans*

Body Length: Up to 108.0 cm.

Like other members of the genus, this slender snake has a long pointed snout. This snout lacks a dermal appendage, separating the species from Long-nosed Whip Snake. The head is not speckled. It is very similar in appearance to the Oriental Whip Snake, with a bright-green, greenish-grey, or grey body, often chequered with pale grey. A white line extends along the outer edge of the ventral scales. It has a single anal shield and 186-195 ventrals and 115-156 sub-caudals; in the very similar Oriental Whip Snake the anal shield is split, the ventrals number 194-235, and the sub-caudals number 151-235. **Behaviour:** Diurnal. An arboreal snake, feeding on birds and lizards. **Habitat:** Occurs in peninsular monsoonal evergreen forest and Malayan mixed dipterocarp forest. **Range:** Peninsular Thailand, south of the Isthmus of Kra. **Regional Range:** Thailand, Peninsular Malaysia, and Indonesia (Java).

254

255 Long-nosed Whip Snake *Ahaetulla nasuta*

Body Length: Up to 190.0 cm.

Distinguished from all other whip snakes by its diagnostic dermal appendage, extending from the end of the snout, which gives a very pointed appearance. The slender body is usually bright green, but coloration can be variable (buff, bluish, brown, grey, pink, yellowish); these colour forms have no geographical distribution. The head is the same colour as the body. A longitudinal yellow line runs along the ventrals. The tail is reddish. The anal is divided. The scales are smooth and in 15 rows at midbody. **Behaviour:** Arboreal in habit. Diurnal. Slowly moves its body back and forth, like a twig moving in the wind, when hunting. Feeds on small mammals, birds, lizards, and frogs. **Habitat:** Found in a wide range of forest types. **Range:** Restricted to central Thailand, south to Prachuap Khiri Khan Province. **Regional Range:** India, Sri Lanka, Bangladesh, Myanmar, Vietnam, Cambodia, and Thailand.

255

256 Speckle-headed Whip Snake *Ahaetulla fasciolata*

Body Length: Up to 140.0 cm.

This whip snake has a characteristic speckled head that is brownish or greyish above, with irregular black speckles and stipples. The body is pale brown or pinkish-grey, often with dark brown crossbars on the neck and the anterior surface of the body. The venter is white, with a black line on the edges of each ventral and sometimes a black median line. **Behaviour:** Arboreal and diurnal. Feeds on lizards and frogs. **Habitat:** Inhabits peninsular monsoonal evergreen forest and Malayan mixed dipterocarp forest. **Range:** Found throughout peninsular Thailand. **Regional Range:** Thailand, Peninsular and Eastern Malaysia, Singapore, Brunei, and Indonesia (Kalimantan and Sumatra).

256

Whip Snakes Genus *Dryophiops*

Members of this group have very slender, compressed bodies. The head is elongate and distinct from the neck. The pupil is horizontal. The frontal is narrow and bell-shaped. A single species, *Dryophiops rubescens*, occurs in Thailand.

257 Brown Whip Snake *Dryophiops rubescens*

Body Length: Up to 64.6 cm.

A small snake with an elongate olive-brownish head that is distinct from its very slender neck and body. The body is usually grey or reddish-brown above with fine flecks and marbling. It has median black dots running irregularly along the dorsal surface. A faint brown stripe runs from the tip of the snout through the eye continuing some distance along the neck. The chin and throat are yellowish. The pupil of the eye is horizontal. The anterior ventral surface is light yellow, which becomes progressively more mottled brown posteriorly. **Behaviour:** Diurnal and entirely arboreal. Feeds on frogs and geckos. It is ovoviviparous. **Habitat:** Inhabits peninsular monsoonal evergreen forest and Malayan mixed dipterocarp forest, up to 500 m in altitude. **Range:** Found in peninsular Thailand, south of the Isthmus of Kra. Extremely rare throughout its range. **Regional Range:** Thailand, Peninsular and Eastern Malaysia, Singapore, Indonesia, and Philippines (Palawan).

257

Keelbacks Genus *Sinonatrix*

This genus was derived from separating closely related species of the very large grouping *Natrix*, occurring in Southeast Asia. They are oviparous. The anal plate is divided. The dorsal scales are in 19 rows at midbody. One species is represented in Thailand.

258 Chinese Keelback ***Sinonatrix percarinata***

Body Length: Up to 94.0 cm.

A medium-sized snake with a dark olive-green or greyish-olive back. The top of the head is also greyish-olive, while the rest of the head is yellowish-white. The sides of the body are marked with dark brown vertical bars, sometimes with light edges. The ventral surface is white anteriorly, with black spots and speckling posteriorly. **Behaviour:** Diurnal, and both aquatic and terrestrial, often found close to water bodies. Diet consists of fish and frogs. **Habitat:** Upper montane forests. **Range:** Restricted to northern and northeastern (west) Thailand. Recorded in Chiang Mai Province and Phu Luang Wildlife Sanctuary in Loei Province. **Regional Range:** India, Myanmar, China, Vietnam, and Thailand.

258

Keelbacks Genus *Xenochrophis*

This genus comprises moderately large, harmless water snakes, which are active during the day. Their diet consists of fish, frogs, and tadpoles. The head is long and narrow. The eyes are large with round pupils. The teeth on the upper jaw become enlarged posteriorly. The body scales are heavily keeled. The ventrals are rounded. The tail is long. They are oviparous. The dorsal scales are in 17-19 rows at midbody. Four species occur in Thailand.

KEY TO THE THAI *XENOCHROPHIS*

1 a	Scale in 17 rows at midbody; scales smooth	*X. punctulatus*
b	Scales in 18 rows at midbody; scales keeled	**2**

2 a	Two black stripes from the eye	**3**
b	No stripes from the eye	*X. trianguligerus*
3 a	Venter yellowish or ivory; outer posterior edge of ventrals greyish or sometimes blackish; diagonal lines from eye absent or very dim; no black diagonal lateral stripes on neck tending to meet its fellow on nape	*X. piscator*
b	Each ventral scale with a well-defined complete transverse stripe; 2 strong diagonal lines from eye, and a lateral nuchal black line	*X. flavolineatus*

Key modified from Smith (1943); Taylor (1965).

259 Common Keelback *Xenochrophis flavipunctatus*

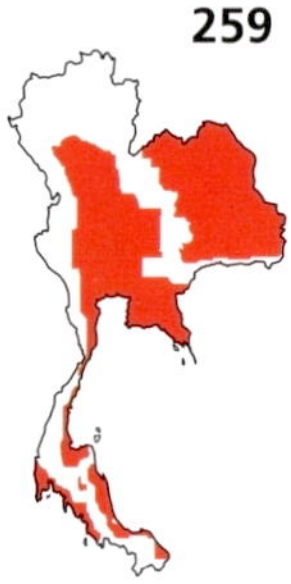

Body Length: Up to 120.0 cm.

A medium-sized robust snake. Head is dark olive, with two distinct black lines running parallel downward and backward from the eyes. The olive body has a row of vertical black bars on the sides, which gradually becomes larger and reticulated in patterning toward the middle of the body, and reduce in size toward the tail. It lacks cream or yellow-cream dorsolateral spots but has very small yellowish marks on scale edges. The venter is greenish white, with each ventral having a narrow black bar. Some individuals show considerable red on the neck. **Behaviour:** Found in ponds, pools, and flooded rice fields. Feeds on fish and frogs. **Range:** Found in low-lying areas throughout the country. **Regional Range:** India, Bangladesh, Myanmar, China, Vietnam, Laos, Cambodia, Thailand, and Peninsular Malaysia.

259

260 **Triangle Keelback** *Xenochrophis trianguligerus*

Body Length: Up to 120.0 cm.

Head and median part of the back is dark olive, with pale-green or yellow throat. Broad olive to olive-brown median stripe, spotted or reticulated with grey-black. The sides of the body have strong black blotches, which are usually triangular; at the apex of each of these blotches, reddish markings may be present. The venter is cream. **Behaviour:** Semi-aquatic habit. Lives in streams, ponds, and other waterbodies. Feeds on frogs. **Habitat:** Found in peninsular monsoonal evergreen forest and Malayan mixed dipterocarp forest. **Range:** Found in peninsular Thailand northward to Kanchanaburi Province. **Regional Range:** India (Nicobar Islands), Myanmar, Vietnam, Laos, Cambodia, Thailand, Peninsular and Eastern Malaysia, Singapore, Brunei, and Indonesia.

260

261 **Chequered Keelback** *Xenochrophis piscator*

Body Length: Up to 84.4 cm.

Large snake often reaching more than 1 m in length. Head is dark green with a black stripe running from the eye to the angle of the jaw, and a second black stripe running from the eye to the supralabial. Olivaceous body with large quadrangular spots arranged in longitudinal and diagonal rows. It has a series of cream or yellow-cream dots covering parts of 3-4 scales extending more than half the length of the body. It lacks distinct black markings on the collar. The chin, venter, and sub-caudal region are yellowish, with the outer edge of the ventrals greyish. **Behaviour:** Both diurnal and nocturnal.

Found in rivers, streams, lakes, marshes and small waterbodies. Feeds on fish and frogs. **Habitat:** Inhabits dry evergreen forest as well as agricultural lands **Range:** Found in northern Thailand. **Regional Range:** Afghanistan, Pakistan, India, Nepal, Bhutan, Sri Lanka, Bangladesh, Myanmar, China, and Thailand.

261

---

262 Burmese Keelback ***Xenochrophis punctulatus***
Body Length: Up to 64.0 cm.

Medium-sized snake with a dark brown upper body and tail; small whitish-yellow dots on the back located on scale-rows 4-5 anteriorly and scale-rows 5-7 farther down the body. The rounded head is dark brown above, being paler on the lateral sides of the snout and cream below. The eyes are small, with rounded pupils. The supralabials are whitish-yellow. The ventrals and sub-caudals are cream, anteriorly and laterally edged with dark brown margins. **Behaviour:** Largely aquatic. **Habitat:** Occurs in mixed deciduous forest. **Range:** Recently recorded from Mae Hong Son Province, northwestern Thailand. **Regional Range:** India, Myanmar, and Thailand.

Mountain Keelbacks Genus *Opisthotropis*

This group is found at high altitudes in forested environs. The head is depressed and the snout broadly rounded. The eye is small or moderate. The maxillary teeth are small, numbering 20-40. The anal is divided. The ventrals are rounded. The dorsal scales are in 17-19 rows at midbody. Three species occur in Thailand.

KEY TO THE THAI *OPISTHOTROPIS*

1 a	Dorsal scles in 17 rows at midbody	**2**
b	Dorsal scales in 19 rows at midbody	*O. praemaxillaris*
2 a	Olive above; pale yellowish below, the colour mingling on 3 outer rows of scales; sub-caudals mottled with grey	*O. spenceri*
b	Body and tail glossy black above, with single yellow spot on each scale; ventrals and sub-caudals yellow with black anterior and lateral margin	*O. maculosus*

Key prepared by Chan-ard, Parr and Nabhitabhata.

263 Spencer's Mountain Keelback ***Opisthotropis spenceri***
Body Length: Up to 60.0 cm.

Slender-bodied snake with an olive body and olive top of the head, while the ventral surface is pale yellowish. The snout is broadly rounded and the small head is distinct. The sub-caudals are mottled with grey. The scales are in 17 rows at midbody. **Behaviour:** Terrestrial. **Habitat:** Occurs in lower and upper montane forests, near forest streams. **Range:** Found in Phrae Province, northern Thailand. **Regional Range:** Endemic to Thailand. **International Conservation Status:** IUCN: Data Deficient.

264 Angel's Mountain Keelback ***Opisthotropis praemaxillaris***
Body Length: Up to 22.7 cm.

A very small snake. The dorsal coloration is uniformly brownish grey. It has indistinct, regularly spaced, pale-brown markings along the flanks. The venter is yellow. The head is barely distinct from the body. The supra-labials are white with brown edges. The ventrals number 145-149 and the

264

sub-caudals number 63-67. The dorsal scales are in 19 rows at midbody. **Behaviour:** Poorly known. **Habitat:** Found in sub-montane forest. Lives near fast-moving mountain streams. **Range:** In Thailand, only known from Doi Saket mountain in Chiang Mai Province, and Nan Province. **Regional Range:** China, Laos, and Thailand.

265 Yellow-spotted Mountain Keelback *Opisthotropis maculosus*

Body Length: Up to 52.0 cm.

The body and top of head are glossy black, becoming lighter brown on the flanks. The ventral surface is creamy-white or mottled brown. The snout is broadly rounded and the head is distinct. The chin is yellow. Sub-caudals mottled with grey. The dorsal scales are in 17 rows. **Behaviour:** Nocturnal and aquatic. **Habitat:** Favours mountain streams in lower and upper montane forest. **Range:** Currently only recorded in Phu Wua Wildlife Sanctuary, Bueng Kan Province, in northeastern Thailand. **Regional Range:** Endemic to Thailand.

265

Hill Stream Keelback Genus *Parahelicops*

Natricine snake with 25-30 maxillary teeth, last two or three are enlarged ungrooved fangs; no apical pits; scales keeled, in 17 or 19 rows, anal double or single; prefrontals fused into a single scale. One species occurs in Thailand.

266 Boonsong's Mountain Keelback *Parahelicops boonsongi*

Body Length: Up to 100.0 cm.

The body and top of head are brown, becoming lighter brown on the flanks. The ventral surface is creamy-white or mottled brown. The snout is broadly rounded and the head is distinct. The sub-caudals mottled with grey. The dorsal scales are in 17 rows. **Behaviour:** Nocturnal and aquatic, feeding on fish and frogs. **Habitat:** Favours mountain streams in lower and upper

266

montane forest. **Range:** Currently only recorded in Phu Luang Wildlife Sanctuary, Loei Province in northeastern (west) Thailand. **Regional Range:** Endemic to Thailand. **International Conservation Status:** IUCN: Data Deficient.

Asian Keelbacks Genus *Rhabdophis*

These snakes are terrestrial and semi-aquatic. Their eyes are rather large and the pupils are round. The scales are heavily keeled. Three species are represented in Thailand.

KEY TO THE THAI *RHABDOPHIS*

1	**a**	Glands present in nuchal area	**2**
	b	Nuchal glands absent; a white chevron on the neck pointing backward	*R. chrysargos*
2	**a**	Olive-green with vertical black bars on sides or narrow crossbars; lips white with two dark oblique stripes	*R. nigrocinctus*
	b	Olive-brown or greenish, uniform or reticulated; head dark, with blackish colour extending on to neck, followed by smallish reddish or vermillion area	*R. subminiatus*

Key modified from Taylor (1965).

267 **Red-necked Keelback** *Rhabdophis subminiatus*

Body Length: Up to 130.0 cm.

A large stout-bodied snake. The head is blackish-brown, with reddish or vermillion coloration on the neck. The body is olive-brown or greenish above. The belly is yellowish, the ventrals often with outer dots. The juvenile has oval blackish spots on the dorsum, a grey head, a black nuchal patch and a black sub-ocular bar. **Behaviour:** Diurnal. Terrestrial, aquatic, and sometimes arboreal. Feeds on amphibians. **Habitat:** Found in dry dipterocarp, mixed deciduous, and dry evergreen forests. **Range:** Found in forested areas throughout the country apart from the extreme southern provinces. **Regional Range:** India, Nepal, Bangladesh, Myanmar, China, Vietnam, Laos, Cambodia, Thailand, Peninsular and Eastern Malaysia, Singapore, and Indonesia.

267

268 Green Keelback *Rhabdophis nigrocinctus*

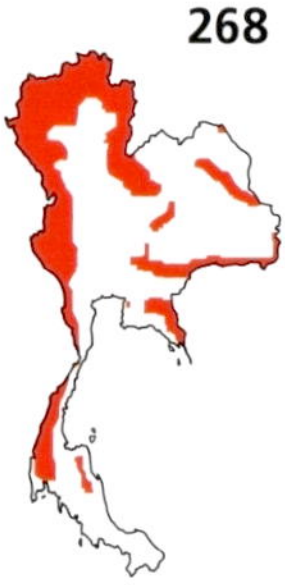

Body Length: Up to 88.0 cm.

The upper body is olive-green anteriorly, becoming brown posteriorly, usually with narrow, vertical black bars. It is uniformly whitish below. The lips have 2 narrow black stripes; one below the eye, the other from behind the eye to the angle of the mouth. The young have a white nape, edged with black in front. The scales are in 19 rows. **Behaviour:** Nocturnal and semi-aquatic. Diet consists of fish and amphibians. **Habitat:** Found in a wide range of evergreen forest types. **Range:** Found throughout the country apart from the extreme southern provinces. **Regional Range:** Myanmar, China, Vietnam, Laos, and Thailand.

268

269 Speckle-bellied Keelback *Rhabdophis chrysargos*

Body Length: Up to 71.5 cm.

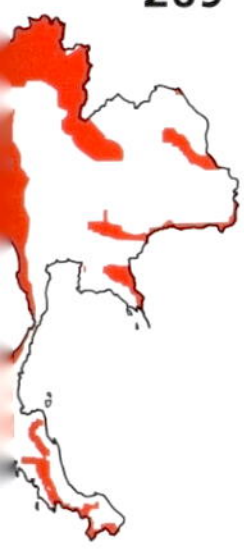

A medium-sized snake. The head is dark green or black, while the lips and throat are strikingly white, and this whitish colour extends backward to form a chevron upon the nape. The body is olive-brownish, greenish, or greyish above with a dorsolateral series of short white or yellow transverse bars, edged with black. The venter is whitish, with black spots on the outer margins of the ventral shields, hence the name "Speckle-bellied". The scale-rows number 19 at the midbody. **Behaviour:** Diurnal. Found along streams. Feeds on rodents, birds, lizards, and frogs. **Habitat:** Found in evergreen, sub-montane and montane forests. **Range:** Found in forested areas throughout the country. **Regional Range:** Myanmar, China, Laos, Cambodia, Thailand, Peninsular and Eastern Malaysia, Brunei, Indonesia, and the Philippines (Palawan).

269

Keelbacks Genus *Amphiesma*

This group is characterized by lateral nostrils, being broadly truncate and having more than 30 maxillary teeth. The genus is represented by 6 species in Thailand.

KEY TO THE THAI *AMPHIESMA*

1 a Posterior maxillary teeth gradually enlarged **2**

b Last two or three maxillary teeth abruptly enlarged; midbody scale-rows 19; internasal much narrowed anteriorly; two light stripes down the back *A. stolatum*

2 a	Eight supralabials; maxillary teeth 20-24; olive-brown or greyish-brown above, with two light distinct dorsolateral black edged stripes along the back and tail	*A. bitaeniatum*
b	Nine supralabials; maxillary teeth 26-36	**3**
3 a	Labial black with light center	*A. khasiense*
b	Labial light with black marginal edges	**4**
4 a	Venter with three series of quadrangular dark or black spots forming lines	*A. deschauenseei*
b	Not so marked	**5**
5 a	Black above with a dorsolateral chain of yellow spots; two outer scale-rows with yellow centers; outer edges of ventrals and sub-caudals with black spots; ventrals, 147-154, sub-caudals, 120-126; anal single	*A. groundwateri*
b	Indefinite grey-to-brown reticulation with two very narrow light dorsolateral cream lines; labial cream, the colour continues as a line diagonally on neck and joining dorsolateral lines; venter yellow, the ventrals with an outer grey or blackish spots	*A. inas*

Key modified from Taylor (1965); Tweedie (1983).

270 Northern Keelback *Amphiesma deschauenseei*

Body Length: Up to 48.0 cm.

Glossy metallic bronze-grey body, with two lighter series of dots or x-shaped spots anteriorly, which grade into a continuous lighter stripe along the length of the body and tail. Between the two lighter stripes are an indefinite series of dark spots. Dorsal scale-rows are keeled. The venter has three series of characteristic quadrangular dark or black spots forming lines. The head is variegated brown. The chin, throat, and supralabials are cream. **Behaviour:** Nocturnal. Terrestrial, associated with forest streams.

270

Habitat: Found in dry dipterocarp forest and mixed deciduous forest. **Range:** Isolated records in Chiang Mai Province, northern Thailand: and Huai Kha Khaeng Wildlife Sanctuary, Uthai Thani Province in western Thailand. **Regional Range:** Endemic to Thailand.

---

271 Groundwater Keelback *Amphiesma groundwateri*

Body Length: Up to 45.0 cm.

Black above with a dorsolateral chain of yellow or white spots, connected by a festooned stripe. Two outer scale-rows with yellow or white centers. Black head has a yellow stripe running from the angle of the mouth to the nape. V-shaped dark mark behind the occiput. Ventrals and sub-caudals are yellowish with the outer edges having black spots. **Behaviour:** Nothing is currently known of its natural history. **Habitat:** Only known from peninsular monsoonal evergreen forest. **Range:** A single record from Tasan, Chumphon Province in peninsular Thailand. **Regional Range:** Endemic to Thailand.

271

---

272 Malayan Mountain Keelback *Amphiesma inas*

Body Length: Up to 50.2 cm.

A medium-sized, slender dark brown keelback. Dark brown dorsum is marked with a dorsolateral row of lighter spots running the length of the body. The flanks have yellow spots. The ventral surface is yellow. The head is brown marked with black. A light cream stripe runs from the snout across the angles of the mouth and converging onto the neck. **Behaviour:** Nocturnal and terrestrial. **Habitat:** Lower montane and upper montane forests. **Range:** Found in Khao Luang National Park, Nakhon Si Thammarat Province in peninsular Thailand. **Regional Range:** Thailand and Peninsular Malaysia.

272

273 White-striped Keelback *Amphiesma stolatum*

Body Length: Up to 73.0 cm.

Distinctly patterned snake, with a contrasting greenish-brown head and an olive-brown or greyish-brown body which has two white, yellowish, or buff dorsolateral stripes running along its entire length. The venter is bluish-white. **Behaviour:** Diurnal and predominately terrestrial. It feeds on fish, lizards, amphibians, and insects. **Habitat:** Occurs in a wide range of forest types, from lowlands up to 2,000 m. Also occurs in grasslands. Usually found in the vicinity of waterbodies. **Range:** Found throughout continental Thailand. Fairly common. **Regional Range:** India (Andamans), Myanmar, China, Vietnam, Laos, Cambodia, Thailand, and Eastern Malaysia.

273

274 Khasi Keelback *Amphiesma khasiense*

Body Length: Up to 60.0 cm.

Dark-greyish or rich-brown above, with or without indistinct light dorsolateral stripes or a series of spots. The ventrals and sub-caudals are yellowish, with the outer margins brown. The top of the head has light vermiculations, and usually two small spots. The labials are white or yellow in the middle, with black borders. The lips are yellow, with the coloration extending backward as a series of spots on each side of the neck. Scales are in 19 rows. **Behaviour:** Diurnal and predominately terrestrial. Closely associated with forest streams. It feeds on frogs, toads, and insects. **Habitat:** Occurs in sub-montane forest at 900-1000 m. **Range:** Found in northern and northeastern (west) Thailand. **Regional Range:** India, Myanmar, China, Vietnam, Laos, and Thailand.

274

275 Striped Keelback *Amphiesma bitaeniatum*

Body Length: Up to 57.0 cm.

Olive-brown or greyish-brown above, the scales sometimes black-edged, and with two light yellow distinct dorsolateral black-edged stripes or series of spots along the back and tail. Supralabials are ivory-yellow. Venter is creamish-yellow. **Behaviour:** Diurnal and predominately terrestrial. It feeds on frogs and toads. **Habitat:** Inhabits upper montane forest. **Range:** Found in Doi Inthanon National Park, Chiang Mai and Nan provinces in northern Thailand. **Regional Range:** Myanmar, China, and Thailand.

275

Bamboo Snakes Genus *Pseudoxenodon*

These snakes are arboreal. The head is distinct from the neck. Their eyes are rather large and the pupils are round. Maxillary teeth number 20-28, increasing in size posteriorly, with the last two greatly enlarged and separated from the others slightly. The body is cylindrical. The scales are keeled and arranged obliquely. The ventrals are rounded. The tail is moderate and the sub-caudals are paired. They are oviparous. One species, *Pseudoxenodon macrops*, occurs in Thailand.

276 Large-eyed Bamboo Snake *Pseudoxenodon macrops*

Body Length: Up to 61.8 cm.

Medium-sized snake with large eyes, with round pupils. The head is greyish-brown with a broad black chevron on the neck preceeded by a narrow orange chevron. The supralabials are white with narrow black edges. The body pattern is highly variable, with a vertebral series of yellowish, reddish-brown or orange, dark-bordered spots. These black spots are comparatively long and narrow anteriorly, but broaden toward the posterior end of the body. The venter is white anteriorly marked with black blotches that grow smaller and less numerous farther along the belly; they are light brown in the posterior third of the belly. The ventrals number 151-180 and the sub-caudals number 55-80. The scale-rows are arranged in 17-19 rows at midbody. **Behaviour:** Terrestrial. Diet consists of lizards and frogs. Threat display involves raising its forebody and flattening its neck; this behavior lends to its other name “mock cobra”. **Habitat:** Inhabits upper and lower montane forest. **Range:** Recorded in northeastern (west) Thailand. **Regional Range:** India, Nepal, Myanmar, China, Vietnam, Laos, Thailand, and Peninsular and Eastern Malaysia.

276

Keelbacks Genus *Macrophistodon*

These snakes are medium-sized with very stout bodies. They feed upon lizards, frogs, and tadpoles. The head is distinct from the body. The eyes are moderately large and the pupils are round. It has 1 pair of very large fangs, hence the name for the genus, "*Macrophistodon*." The dorsal scales are strongly keeled (hence the name "keelback"), with apical pits, and are arranged in 19-27 rows at midbody. The ventral are rounded. The tail is moderately long. The sub-caudals are paired. They are oviparous. Two species are present in Thailand.

KEY TO THE THAI *MACROPHISTODON*

1 a Body greyish-rose or reddish brown with a dark brown or black vertebral stripe which becomes a broad V-shaped patch on the neck; sides of venter pink, middle part yellow; 7 or 8 upper labials; maxillary teeth 14-17 — *M. rhodomelas*

b Body black; a broad orange band across the neck separated from head by a broad black band; belly olive with black crossbands, or entirely black; 8 upper labials normally; maxillary teeth 20-22 — *M. flaviceps*

Key modified from Taylor (1965).

277 Blue-necked Keelback *Macrophistodon rhodomelas*

Body Length: Up to 60.0 cm.

A very small snake. The dorsal body is reddish-brown or greyish-rose with a narrow brown or black dorsal stripe. The top of the head is reddish-brown, while the neck has a broad dark brown or black chevron; this coloration contrasts with the sides of the neck which are characteristically pale blue.

The sides of the belly are pink. The venter is mainly yellow, with the tail whitish below. The ventrals number 124-138 and the sub-caudals number 42-58. **Behaviour:** Terrestrial and semi-aquatic. Active at day and night. Feeds on lizards, frogs, and tadpoles. It can dilate its neck in the same manner as a cobra. **Habitat:** Found in wetland areas in lowlands, in plantations, peninsular monsoonal evergreen forest, and Malayan mixed dipterocarp forest. **Range:** Found throughout peninsular Thailand. **Regional Range:** Thailand, Peninsular and Eastern Malaysia, Singapore, and Indonesia.

277

278 Red-headed Keelback *Macrophistodon flaviceps*

Body Length: Up to 84.0 cm.

A medium-sized, stout-bodied snake. The head is reddish dorsally with a broad orange band across the neck, in front of which is a black band. The body is black with greyish-white or pale olive crossbars. The dorsal scales are very heavily keeled in 19 rows. The belly is either olive with black cross-bands, or entirely black. The ventrals number 120-138, and the sub-caudals number 49-60. **Behaviour:** Diurnal. Semi-terrestrial and semi-aquatic.

278

Found in wetlands within forests. Diet includes lizards and frogs. **Habitat:** Occurs in peninsular monsoonal evergreen forest and Malayan mixed dipterocarp forest. **Range:** Found in Peninsular Thailand. **Regional Range:** Thailand, Peninsular and Eastern Malaysia, and Indonesia (Kalimantan and Sumatra).

Homalopsine Water Snakes

Nostrils valvular, on upper surface of snout. Maxillary teeth decreasing in length posteriorly followed by grooved poison fangs. Tail not compressed. (This subfamily has been reclassified and is now the Family Homalopsidae).

KEY TO THE THAI HOMALOPSIDAE

1	**a**	Rostral appendages present. Ventral scales only slightly larger than dorsal scales	*Erpeton*
	b	No rostral appendages. Most ventral scales wider than dorsal scales	**2**
2	**a**	Nasal scales in contact	**3**
	b	Nasal scales separated by one or more internasal scales	**5**
3	**a**	Dorsal scales smooth, at least on the anterior of body	*Enhydris*
	b	Dorsal scales keeled	**4**
4	**a**	Parietal scales fragmented, dorsal scales in 23-29 rows at midbody	*Cerberus*
	b	Parietal scales entire. Dorsal scales in 33-46 rows at midbody	*Homalopsis*
5	**a**	Dorsal scales triangular and in 37-43 rows. Ventrals keeled	*Bitia*
	b	Dorsal scales in fewer than 37 rows, ventral scales not keeled	**6**
6	**a**	Loreal absent. 5-7 upper labials, 23-27 dorsal rows at midbody	*Fordonia*
	b	Loreal present	**7**
7	**a**	Dorsal scales in 17 rows, prefrontals are almost round	*Gerada*
	b	Dorsal scales in 19-21 rows at midbody, lower labials 7-8	*Cantoria*

Key modified from Murphy (2007)

Mud Snakes Genus *Enhydris*

All members of this group have smooth scales. They have cylindrical bodies and partially flattened heads that are often indistinct from the body. They have small eyes with vertical pupils. The tails are strongly depressed. Six species occur in Thailand.

KEY TO THE THAI *ENHYDRIS*

1	**a**	Scales in 19 rows; olive to olive black above; Two outer scale-rows usually yellow; yellow or white belly	*E. plumbea*
	b	Scales in 21-29 rows	**2**

2 a Scales in 27 or 29 rows; greenish-black to olive with yellowish transverse bars or series of spots; below yellow, dark colour of back tending to form rings across venter; loreal touching internasal *E. bocourti*

b Scales in 21 rows at midbody **3**

3 a Lateral-dorsal pattern contains stripes or stripes with spots; anterior chin shields are usually long and narrow. The first labial usually contacts the loreal scale **4**

b Lateral-dorsal pattern consists of large, scalloped, dark blotches that arise from the ventral side and narrow dorsally; or, small scattered spots on a grey ground color, but no dorsal stripes. Anterior chin shields tend to be wide and flared, almost round. Posterior dorsal scales in the first few rows are ovate, almost half-circle in shape **5**

4 a 25-27 scale-rows on neck, 21 (rarely 23) at midbody, anterior chin shields shorter than posterior pair, each nasal scale has a smaller diameter than the eye diameter, more than 153 ventral scales, no lateral spots *E. enhydris*

b 23 (rarely 25) scale-rows on the neck, chin shields variable in size and number, each nasal scale has a diameter greater than eye diameter, 153 or fewer ventral scales, lateral spots above scale-row 4. Dorsal scales on posterior of body in the first few rows are lanceolate and almost square *E. subtaeniata*

5 a Usually 21 scale-rows on neck (rarely 23). Temporal scales distinct and plate-like. Small spots involving three or four scales, outer edge of each ventral involves a zigzag stripe. Ventral scales number 116-125. No large scalloped lateral blotches. Eye diameter equal to nasal diameter or slightly less *E. chanardi*

b 23-25 scale-rows on neck, large scalloped blotches or spots on sides and dorsum. Ventral scales number 117-127. Eye diameter less than diameter of nasal scale *E. jagorii*

Key modified from Taylor (1965); Murphy (2007).

279 Boie's Mud Snake *Enhydris plumbea*

Body Length: Up to 53.0 cm.

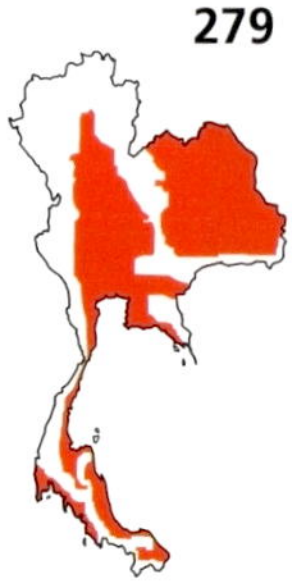

A small water snake reaching just over 50 cm in length. The body is short and stout, being grey, or olive above. The blunt head is uniformly olive. The snout is broadly rounded and is indistinct. The venter is yellow or dull white with a broken row of blackish spots along the median line of the venter. The scales are in 19 rows at the midbody. **Behaviour:** Aquatic. Both diurnal and nocturnal. Feeds on fish, frogs, and crustaceans. **Habitat:** Found in freshwater habitats up to 1,200 m. **Range:** The species is widespread throughout the country. **Regional Range:** Myanmar, China, Vietnam, Laos, Cambodia, Thailand, Peninsular and Eastern Malaysia, and Indonesia.

279

280 **Bocourt's Mud Snake** *Enhydris bocourti*

Body Length: Up to 114.0 cm.

This water snake is the largest member of this genus, reaching more than a metre in length, with a notably stout body. It has a very dark green, almost black body with an irregular series of narrow, transverse, yellowish bars, with each bar bordered by black. The lower sides and chin are yellow. The depressed head is uniformly olive above. Venter is yellow with dark dorsal markings. The scales are in 27 or 29 rows. The ventrals number 124-136. The anal plate is divided. **Behaviour:** Nocturnal and normally aquatic in nature, though found on land in wet conditions. **Habitat:** Found in large lakes and other low-lying freshwater waterbodies. **Range:** Recorded from central plains and lowlands of peninsular Thailand. **Regional Range:** Vietnam, Cambodia, Thailand, and Peninsular Malaysia.

280

281 **Rainbow Mud Snake** *Enhydris enhydris*

Body Length: Up to 74.1 cm.

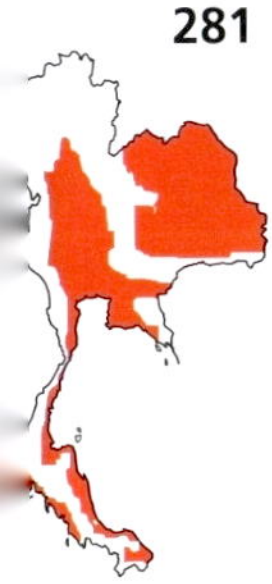

A large, smooth-bodied water snake with an indistinct head, reaching more than 80 cm in length. It has darker and lighter longitudinal stripes, varying in width and colour, running the length of the body. On the dorsal surface, it has a median dark brown stripe, bordered by a light brown stripe. The lower sides (the last three rows of costal scales) are alternately lemon yellow, pale pinkish, and again yellow running the length of the body, hence the "rainbow" name. The ventrals are diagnostically pale yellowish or white with a black line along each side of the ventrals and usually a thin dark median line. **Behaviour:** Diurnal and most usually aquatic. Found in all low-lying freshwater wetlands. It is piscivorous. Often lies quietly among the weeds, with its snout sticking above the water. **Habitat:** Low-lying wetlands, including agricultural fields. **Range:** Found throughout the country. **Regional Range:** Pakistan, India, Nepal, Bangladesh, Myanmar, Vietnam, Laos, Cambodia, Thailand, Peninsular and Eastern Malaysia, Singapore, and Indonesia.

281

282 Mekong Mud Snake *Enhydris subtaeniata*

Body Length: Up to 87.0 cm.

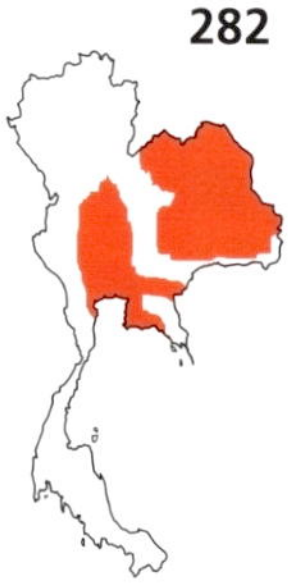

A large, smooth-bodied mud snake with an indistinct head; the largest female reaches more than 80 cm in length and the largest male reaches 65 cm in length. It is dark brown on the outer edge of the ventral scales; combined with the dark-brown pigment on the anterior edge of the first scale-row it forms a zigzag stripe on each side of the belly. The posterior half of the first scale-row is cream. The second scale-row and sometimes the bottom portion of scale-row 3 have a salmon-pink or red stripe with some brown pigment. Scale-rows 4-18 are solid dark brown. Starting behind the parietal scales, these become indistinct posteriorly. The midline of the ventral scales frequently has a series of spots that form a midline stripe on the posterior of the body; this is rarely absent, and rarely does it extend to the anterior quarter of the ventral scales. There is a dark stripe on the midline of the tail bordered by cream stripes above and below. **Behaviour:** Nocturnal and most usually aquatic, though may venture on land in very wet conditions.

282

Usually found in shallow water bodies with aquatic plants. Diet consists of fish. **Habitat:** Low-lying wetlands. **Range:** Found throughout the Mekong Basin, in lowlands and the central plains. **Regional Range:** Vietnam, Laos, Cambodia, and Thailand.

283 Bangkok Mud Snake *Enhydris jagorii*

Body Length: Up to 51.5 cm.

A small mud snake. Above generally greyish-brown. This species can be distinguished from most other species of mud snakess by having 21-23 scale-rows at midbody, by its low ventral scale counts (117-127), and by its sub-caudal scale counts (48-54 in females and 53-68 in males). The anterior pair of chin shields is longer than the second pair (or rarely the same length) and they are flared. It has a lateral-dorsal pattern of dark blotches (about 5-scale-rows long) running from the sides of the neck to the tail. **Behaviour:** Aquatic. **Habitat:** Occurs in waterbodies and marshes with muddy substrates. Feeds on fish and amphibians. **Range:** Recorded from upper and lower central plains in central Thailand. **Regional Range:** Vietnam, Laos, Cambodia, and Thailand.

283

284 Chanard's Mud Snake *Enhydris chanardi*

Body Length: Up to 54.7 cm.

This mud snake has a grey-brown dorsum. It has yellow stripes bordering a lower grey stripe. The chin is uniformly yellow. Rows 5-7 have black spots. Chanard's Mud Snake can be distinguished from all other mud snakess that have 21 scale-rows at midbody by the following characteristics: 116-125 ventrals, 38-60 sub-caudals (53-60 in males and 38-49 in females); two pairs of chin shields with the second longer in length; interocular distance that is greater than the distance of the frontal; scales that are longer than taller on rows 7-15; and a stripe on scale-row 3. **Behaviour:** Little is known about the food habits and reproduction of this species. **Habitat:** Freshwater environs of metropolitan Bangkok and the Inner Gulf. **Range:** Endemic to the Gulf of Thailand and Bangkok. **Regional Range:** Endemic to Thailand.

Water Snakes Genus *Homalopsis*

All members of this group have strongly keeled scales. They have cylindrical bodies which are distinct from the head and neck. They have small eyes with vertical elliptical pupils. The parietals are whole. Two species occur in Thailand.

KEY TO THE THAI *HOMALOPSIS*

1 a	Belly is usually white with dark spots near the edges of each ventral. First divided upper labial usually 7-8	*H. buccata*
b	Belly is black or black-brown with light color spots near the edges of the ventrals. First divided upper labial usually six	*H. nigroventralis*

Key modified from Murphy (2007).

285 Puff-faced Water Snake *Homalopsis buccata*

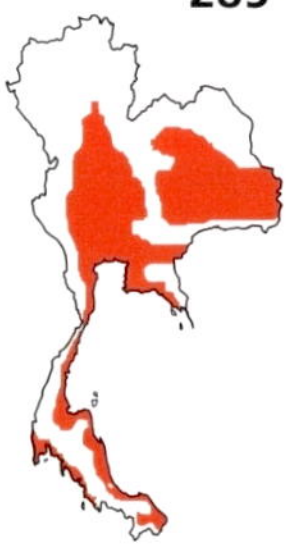

Body Length: Up to 94.4 cm.

This water snake has a large, broadly rounded head. It has a dark brown body with black-edged, grey-brown crossbars. The dorsal surface has a number of irregular mid-brown blotches. Head has a broad chevron on the occiput, as well as V-shaped white mark on the snout. The young are brownish-black above with 27 narrow, ivory-white, transverse bands around the body. **Behaviour:** Nocturnal. Semi-aquatic. Eats fish and frogs. **Habitat:** Inhabits estuaries. Also occurs in canals and fish ponds where it can often be seen fishing in the water or burrowing into the mud. **Range:** Found in freshwater habitats and estuaries throughout Thailand. **Regional Range:** India, Nepal, Bangladesh, Myanmar, Vietnam, Laos, Cambodia, Thailand, Peninsular and Eastern Malaysia, Singapore, and Indonesia.

285

286 Deuve's Water Snake ***Homalopsis nigroventralis***

Body Length: Up to 103 cm.

This water snake has a large, depressed head. It has reddish-brown coloration with the typical banded pattern as seen in the Puff-faced Water Snake. However, they have an irregular cream line that runs from the angle of the mouth posteriorly along the 3rd and 4th scale-rows, this becomes broken into spots. Additionally there is a cream line from the nasal scale, below the eye, to the angle of the mouth. The occipital region has a pair of large cream spots. The ventral pattern is black with lateral cream spots and some anterior transverse bars. **Behaviour:** Both diurnal and nocturnal. Aquatic. Diet consists of fish, amphibians, and crustaceans. **Habitat:** Freshwater wetlands, including rivers, canals, and flooded forests. **Range:** Nakhon Phanom and Sakhon Nakhon provinces in northeast Thailand. **Regional Range:** Laos and Thailand.

286

Bockadam Genus *Cerberus*

Members of this group have broadly rounded snouts. The eye is small with a vertical pupil. The maxillary teeth number 12-17. The parietals are broken up into small scales. They have keeled scales, in 21-29 rows. One species is represented in Thailand.

287 Asian Bockadam ***Cerberus rynchops***

Body Length: Up to 47.0 cm.

The body is more-or-less cylindrical. Above, the dorsum is mainly olivaceous grey with a series of about 65 indefinite black crossbars. The ventrals are olive-grey, with considerable yellow on outer part of ventrals. The frontal is small. The chin is flecked with dark pigment. There is an indistinct black line from the side of the snout through the eye which extends along the neck for some distance. The scales are in 21-29 rows. **Behaviour:** Both nocturnal and diurnal. Feeds on fish including mudskippers and gobies as well as frogs and crustaceans. **Habitat:** Prefers coastal waters where it is common in the soft mud of mangroves. Also found in estuaries and occasionally along waterways farther inland. **Range:** Found on all Thai coasts. **Regional Range:** Pakistan, India, Sri Lanka, Bangladesh, Myanmar, Vietnam, Cambodia, Thailand, Peninsular and Eastern Malaysia, Singapore, Indonesia, the Philippines, and East Timor. **International Conservation Status:** CITES: Appendix III.

287

Water Snakes Genus *Gerarda*

All members of this group have short bodies with a flattened head indistinct from the body. The tail is also short. The scales are in 17 rows. The genus is represented by a single species, which is found in Thailand.

288 Gerard's Water Snake *Gerarda prevostiana*

Body Length: Up to 34.6 cm.

A very small water snake with a cylindrical body reaching just over 30 cm in length. It is uniformly olive-grey above. The three outer scale-rows are ivory-yellow. The chin is yellow. The venter is grey, with each scale having some ivory or ivory-grey on the ends. The sub-caudals are uniformly dark grey. The eyes are very reduced. **Behaviour:** Mainly nocturnal. Occurs in brackish waters, where it consumes fish, shrimps, and other aquatic animals. **Habitat:** Inhabits mangrove forest and coastal waters. **Range:** Occurs in the Inner Gulf of Thailand; at Ang Sila, in Chonburi Province, it is reportedly common. **Regional Range:** India, Sri Lanka, Bangladesh, Myanmar, Thailand, Singapore, and Indonesia (Sumatra).

288

Crab-Eating Snakes Genus *Fordonia*

All members of this group have smooth scales. They have cylindrical bodies and partially flattened heads that are often indistinct from the body. They have small eyes with vertical pupils. One species is found in Thailand.

289 Crab-eating Water Snake *Fordonia leucobalia*

Body Length: Up to 65.0 cm.

A small water snake. There are several colour variations ranging from olive-brown or brown, as well as grey with black spots, to purplish black. In all varieties the dorsal coloration fades to ivory white on the outer four or five scale-rows. The ventrals, throat, and chin belly are ivory-white. It has five upper labials. The jaw is recessed under the head, and the eyes positioned toward the top of the head. It has scales in 25-29 rows. **Behaviour:** Nocturnal and aquatic. Hides in crab burrows. Preys on small fish and crabs. Burrows into the mud or enters crab burrows. **Habitat:** Tidal habitats, particularly mudflats and mangroves. **Range:** Recorded in the Andaman Sea from Phuket to Satun provinces. **Regional Range:** India, Bangladesh, Myanmar, Vietnam, Cambodia, Thailand, Peninsular and Eastern Malaysia, Singapore, Indonesia, the Philippines (Mindanao), Papua New Guinea, and Australia.

289

Mangrove Snakes Genus *Cantoria*

This genus has elongate, cylindrical bodies with heads which are indistinct from the neck. The head has large, distinct, shields. The eyes are small with vertical pupils. The scales are smooth, in 19 rows. The ventrals are moderate or well-developed, lacking keels. The tail is moderate and slightly compressed. The sub-caudals are paired. One species occurs in Thailand.

290 Cantor's Mangrove Snake *Cantoria violacea*

Body Length: Up to 94 cm.

Body is dorsally blackish-brown with 89 transverse pale yellow bands, with 17 found on the tail. These pale yellow bands are darker, brownish-yellow centrally and less distinct on the posterior body compared to the anterior and the tail. The crown of the head is black with irregular sandy-yellow blotches. The upper lip is pale yellow. The outer three scale-rows and the venter are creamy buff. The sub-caudal scales are strongly edged with black, forming a conspicuous central black line down the basal part of the under tail. The dorsal scales are in 19 rows. **Behaviour:** Feeds on the giant mangrove snapping shrimp *Alpheus microrhychus*. **Habitat:** Occurs in tidal mudflats. **Range:** Recorded from Phuket, peninsular Thailand. **Regional Range:** India (Andamans), Myanmar, Thailand, Peninsular Malaysia, Singapore, Indonesia (Kalimantan and Sumatra), and Australia.

290

Estuarine Snakes Genus *Bitia*

All members of this group have smooth body scales with a lateral keel. They have 37-47 scale-rows. The scales on the sides are largely separated from each other by skin. They have very small eyes. A single species occurs in Thailand.

291 Keel-bellied Estuarine Snake *Bitia hydroides*

Body Length: Up to 45.0 cm.

A small water snake with an indistinct head. The body is pale grey above, with black crossbands which extend down just over halfway down the sides of the body. The head is almost uniformly grey or dark brown, with a pale yellow chin and throat. The eyes are located dorsally. The venter and sub-caudal region is pale yellow. The ventrals are narrow with two keels. The tail is short and weakly depressed. **Behaviour:** Aquatic. Diet consists of gobies. **Habitat:** Inhabits estuaries, coastal areas, and mudflats. **Range:** Only recorded in the Inner Gulf and Bangkok. However, records in Myanmar and West Malaysia suggest the species is present in peninsular Thailand. **Regional Range:** India (Andamans), Myanmar, Thailand, Peninsular Malaysia, Singapore, and Indonesia (Sumatra).

291

Tentacled Snake Genus *Erpeton*

Members of this group have a pair of soft appendages growing from the rostrum. They have keeled scales. The ventrals are rather narrow and ventrally angulate. One species occurs in Thailand.

292 Tentacled Snake *Erpeton tentaculatum*

Body Length: Up to 44.5 cm.

Small water snake with diagnostic tentacular appendages on the rostrum. The body is light brown above. The neck has paired spots that join farther down the body to form two narrow blackish lines, which become fragmented toward the tail region. It also has a dark brown lateral stripe. The venter is yellowish with two rows of irregular broken dark spots or flecks extending nearly to the tail. **Behaviour:** Entirely aquatic. Diet consists of fish. **Habitat:** Found in ponds and sluggish water. **Range:** Recorded in central plains, as well as Thale Noi, Phatthalung Province and Songkhla Lake, Nakhon Si Thammarat Province. **Regional Range:** Vietnam, Cambodia, and Thailand.

292

Elaphid Snakes Family Elapidae

This family comprises the kraits, cobras, and some coral snakes. These snakes have grooved poison fangs at the front end of the maxillary bone. All members are poisonous. Twelve species are found in Thailand.

KEY TO THE THAI ELAPIDAE

1 a	Very large snakes (to 5.5 m in length) with a pair of large postparietal shields; poison glands are on sides of head, poison fangs followed usually by three solid maxillary teeth; 15 scale-rows at midbody	*Ophiophagus*
b	Rather large (nearly 2 m), medium or small snakes lacking enlarged postparietal shields	**2**
2 a	Vertebral series of scales enlarged	*Bungarus*
b	Vertebral series of scales not enlarged	**3**
3 a	Scales in 13 rows, no elongated nuchal ribs permitting the nuchal skin to be stretch into a hood. Transverse rows of scales not arranged obliquely	**4**
b	Scales in 15-21 rows, with elongated nuchal ribs permitting the nuchal skin to be spread into a hood. Transverse scale-rows obliquely arranged	*Naja*

4 a	Anal shield single or divided. Six supralabials	*Calliophis*
b	Anal shield divided. Seven supralabials	*Sinomicrurus*

Key modified from Taylor (1965); Slowinski et al. (2001)

Kraits Genus *Bungarus*

This group comprises medium-to-large terrestrial snakes. Kraits are nocturnal, being often docile by day and becoming aggressive at night. They consume cold-blooded prey, being mainly snake eaters but several species eat lizards. They have potent venom. Three species occur in Thailand.

KEY TO THE THAI *BUNGARUS*

1 a	Scale-rows number 13 at midbody; black with an orange-yellow or red median dorsal stripe; and a red or orange head	*B. flaviceps*
b	Scales-rows number 15, rarely 17; sub-caudal scales single throughout	**2**
2 a	Tail ending in a point; dorsal vertebrae do not form a ridge down back; dorsum black with 20-25 broad white cross-bars, the median part of each bar spotted with black; head black above	*B. candidus*
b	Tail ending bluntly, often with a swollen tip; has a mid-dorsal ridge on body and tail; body alternately banded black and yellow; a large black mark on nape reaches to between the eyes	*B. fasciatus*

Key modified from Taylor (1965).

293 **Red-headed Krait** *Bungarus flaviceps*

Body Length: Up to 185.5 cm.

A large snake. The body is iridescent black above, with a distinct ridge in the middle of the back and tail, though less distinct than in Banded Krait. The head, neck, and tail are red, orange, or yellowish. The black eyes contrast strongly with the head coloration. Evenly spaced, small white spots are present on the vertebral ridge and along the sides of the body. It has a lateral stripe of reddish-orange or yellow. The venter is usually brown or blackish, becoming lighter before the vent; farther back this becomes red. The tail is tipped with a small black spot. **Behaviour:** Nocturnal and terrestrial. Diet consists of other snakes as well as lizards. **Habitat:** Inhabits peninsular monsoonal evergreen forest and Malayan mixed dipterocarp forest. **Range:** Found throughout peninsular Thailand as far north as Petchaburi Province. **Regional Range:** Myanmar, Vietnam, Cambodia, Thailand, Peninsular and Eastern Malaysia, and Indonesia (Kalimantan and Sumatra).

293

294 Blue Krait *Bungarus candidus*

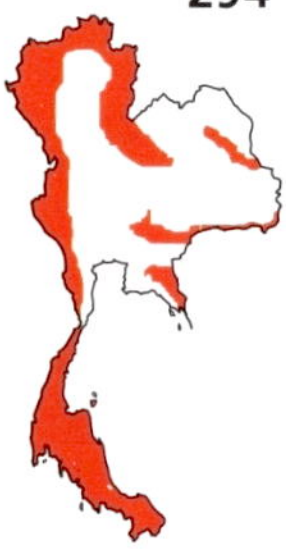

Body Length: Up to 130.0 cm.

The smallest of the krait species, but still a sizable snake, reaching 130 cm. The dorsum is characteristically white or grey, with iridescent, saddle-like, bluish-black bands alternating along the length of the body. The black bands terminate at the ventral scales, while the white bands fuse with the ivory-white venter. The top of the head is greyish-black, followed by a chevron-shaped lighter area. The sub-caudals are single. The tail ends in a point. The ventrals range from 194 to 237 and the sub-caudals number from 37 to 56. **Behaviour:** Nocturnal and terrestrial. **Habitat:** Found in dry evergreen, southeastern and peninsular monsoonal evergreen, and Malayan mixed dipterocarp forests. **Range:** Found in forested areas throughout the country. **Regional Range:** Vietnam, Cambodia, Thailand, Peninsular Malaysia, Singapore, and Indonesia.

294

295 Banded Krait *Bungarus fasciatus*

Body Length: Up to 212.5 cm.

A very large snake. Generally yellow or buff with 28 evenly-spaced black bands encircling the entire body. It has a prominent ridge on the back and tail. It has a large elongate mark on the neck extending forward to between the eyes. The supralabials largely white or yellow. The chin is immaculate.

Scales in 15 rows throughout. The tail is blunt, being swollen at the tip. The venom is highly neurotoxic. **Behaviour:** Terrestrial. Active by night, being docile during the daytime. Feeds on other vertebrates, including rodents, lizards, and other snakes. **Habitat:** Found in a wide range of open forest habitats and wet areas, usually at low elevations. **Range:** Found throughout Thailand, in the lowlands. **Regional Range:** India, Nepal, Bhutan, Bangladesh, Myanmar, China, Vietnam, Laos, Cambodia, Thailand, Peninsular and Eastern Malaysia, Singapore, Brunei, and Indonesia.

295

Cobras Genus *Naja*

This group is characterized by the elongate nuchal ribs being able to expand the anterior part of the neck into a hood. In Thailand, the genus is represented by three species.

KEY TO THE THAI *NAJA*

1	**a**	15-19 scale-rows at midbody. The hood without marking	*N. sumatrana*
	b	19-21 (usually 21) scale-rows at midbody. The hood has a marking	**2**
2	**a**	The hood with a single monocle or O-shaped cream mark	*N. kaouthia*
	b	The hood mark is V, U, or, most commonly, spectacle-shaped	*N. siamensis*

Key prepared by Chan-ard, Parr and Nabhitabhata.

296 Monocled Cobra *Naja kaouthia*

Body Length: Up to 200.0 cm.

A large snake with a hood with a single monocle- or O-shaped cream mark. The chin and throat are ivory. The throat is pale, with scarcely any dark mottling, often followed by a single dark band. The ventro-lateral throat spots are distinct. The remainder of the venter is either pale or increasingly cloudy with darker pigmentation toward the rear. In adults, the hood markings are usually distinct, usually a pale oval or circular marking, with a dark center and occasionally a narrow dark outer border; occasionally one or two dark spots are present in the pale oval. Fangs not modified for spitting, venom discharge orifice large. The body scales are smooth, arranged in 19-21 (usually 21) longitudinal rows at midbody. The ventrals number 164-196 and the sub-caudals 43-58. **Behaviour:** Crepuscular and terrestrial. Diet consists of rodents, fish, other snakes, and frogs. **Habitat:** Found in forests, agricultural areas including rice paddies, and plantations up to 800 m. **Range:** Found throughout the country. **Regional Range:** India, Nepal, Bhutan, Bangladesh, Myanmar, China, Vietnam, Laos, Cambodia, Thailand, and Peninsular Malaysia. **International Conservation Status:** CITES: Appendix II.

296

---

297 Indochinese Spitting Cobra ***Naja siamensis***

Body Length: Up to 160.0 cm.

A large snake with a highly variable body pattern. Individuals from northern and eastern Thailand tend to be uniformly light brown, olive, or sometimes distinctly greenish, often with a somewhat faded appearance. In central Thailand, some have a very contrasting black-and-white pattern, with or without speckling and cross-banding, and a light venter with or without broad dark crossband; others are some shade of brown or greyish-brown, with or without lighter crossbands on dorsum, often with several broad dark bands across belly. Others, especially from the west, are uniformly black. The hood mark in northern, eastern, and southeastern Thailand is V-, U-, or most commonly, spectacle-shaped, but often very indistinct or absent altogether. In central Thailand, H-shaped hood marks are also common, but the hood marks are often absent altogether. There are 25-31 scale-rows around hood, and 19-21 just ahead of the midbody. It has 153-174 ventrals and 45-54 sub-caudals. The basal pairs are sometimes undivided. **Behaviour:** Nocturnal and terrestrial. Feeds on rodents. **Habitat:** Found in a wide range of forest habitats in lowlands. **Range:** Found in western, central, and northeastern Thailand. **Regional Range:** Vietnam, Laos, Cambodia, and Thailand. **International Conservation Status:** IUCN: Lower Risk—Near Threatened; CITES: Appendix II.

297

298 **Sumatran Spitting Cobra** *Naja sumatrana*

Body Length: Up to 150.0 cm.

Uniformly yellowish or light brown, with or without a clear pattern on the throat. There is no hood pattern. It has 19-27 rows around the hood (normally 21-25), and 15-19 just ahead of midbody; 179-201 ventrals, 40-57 sub-caudals. The basal sub-caudals are often undivided. **Behaviour:** Nocturnal and terrestrial. Feeds on rodents and other small vertebrates **Habitat:** Found in lowlands in peninsular monsoonal evergreen forest and Malayan mixed dipterocarp forest. **Range:** Peninsular Thailand, south of the Isthmus of Kra, in forested areas. **Regional Range:** Thailand, Peninsular and Eastern Malaysia, Singapore, Brunei, Indonesia and the Philippines (Palawan). **International Conservation Status:** CITES: Appendix II.

298

King Cobra Genus *Ophiophagus*

This group is represented by a single species, the King Cobra, the largest poisonous snake in the world. It feeds on cold-blooded animals, particularly other snakes. It is oviparous.

299 King Cobra *Ophiophagus hannah*

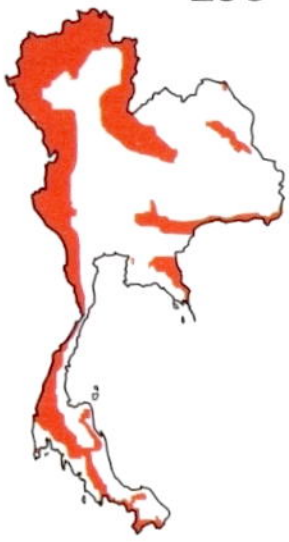

Body Length: Over 5 m.

A very large, generally olive, snake capable of reaching a length of more than 5 m. The head bears diagnostic large shields. The neck is slightly dilatable. The chin and the throat are yellowish. The ventrals are grey. Anterior sub-caudals are single. Scales in 15 rows at midbody. The young show strongly contrasting black-and-yellow bands. **Behaviour:** Diurnal and terrestrial. It preys almost exclusively on other snakes, and requires a large territory. It is noted for its habit of building a nest with sticks and twigs. **Habitat:** Found in most forest types. **Range:** Found in forested areas throughout the country. **Regional Range:** Pakistan, India, Nepal, Bhutan, Bangladesh, Myanmar, China, Vietnam, Laos, Cambodia, Thailand, Peninsular and Eastern Malaysia, Singapore, Brunei, Indonesia, and the Philippines. **International Conservation Status:** CITES: Appendix II.

299

Long-Glanded Coral Snakes Genus *Calliophis*

These snakes are small to medium-sized. All species have an elongate poison gland extending into the anterior third of the body, hence their name of Long-glanded Coral Snakes. They are very slender, with small heads. The eyes are small with round pupils. The ventrals are rounded. When they feel threatened, these ground-dwelling snakes curl up their tails laterally to form a flat spiral, which is then displayed with the colourful underside up. The group is represented by 4 species.

KEY TO THE THAI *CALLIOPHIS* AND *SINOMICRURUS*

1	**a**	Anal scale divided	**2**
	b	Anal scale single	**4**
2	**a**	Pattern not characterized by small black spots on dorsum with two black bands or rings on tail	**3**
	b	Pattern characterized by small black spots on dorsum, sometimes forming stripes, with two black bands or rings on tail, one at the base and one at the tip	*Callophis maculiceps*
3	**a**	Six supralabials	*C.s gracilis*
	b	Seven supralabials	*Sinomicrurus macclelandi*
4	**a**	Snout and venter free of dark pigment	*C. bivirgata*
	b	Snout and venter with dark pigment	*C. intestinalis*

Key modified from Slowinski et al. (2001).

300 Spotted Coral Snake *Calliophis gracilis*

Body Length: Up to 35.5 cm.

A very slender small snake. Grey to brownish-grey body colour. A narrow median black stripe runs from the head to the tip of the tail. A pair of large lateral nuchal spots followed by a series of 25 lateral, paired, black spots terminating at the vent, as well as a median series of smaller spots interrupting the median stripe. The venter is black with white bars. The ventral side of the tail is pink, which is shown when it feels threatened. **Behaviour:** Ground-dwelling and nocturnal. Diet consists of other snakes. **Habitat:** Inhabits Malayan mixed dipterocarp forest at low altitudes. **Range:** Found in peninsular Thailand. Recorded in Na Pradoo, Pattani Province. **Regional Range:** Thailand, West Malaysia, and Indonesia (Sumatra).

300

301 Blue Long-glanded Coral Snake *Calliophis bivirgata*

Body Length: Up to 139.9 cm.

An extremely slender snake, with a striking coral red head and short tail; the coloration on the head extending back on the sides of the neck. Around the eyes is some bluish-black coloration. The slightly iridescent body is blackish-blue above; with a narrow greyish or ultramarine-blue line on the lower sides extending along the entire length of the body. The venter is immaculate red. It has an elongated poison gland. It has smooth scales, in 13 rows. **Behaviour:** Terrestrial. It feeds on small rodents, lizards, and frogs. Threat display involves concealing its head under the coils of its body and exposing its brightly coloured tail. **Habitat:** Inhabits Malayan mixed dipterocarp forest. **Range:** Found in extreme southern Thailand. Recorded from Songkhla, Pattani, Yala, and Narathiwat provinces. **Regional Range:** Cambodia, Thailand, Peninsular Malaysia, Singapore, and Indonesia.

301

302 Brown Long-glanded Coral Snake *Calliophis intestinalis*

Body Length: Up to 58.0 cm.

A small, slender snake, usually reaching about 0.5 m in length. The brown body is diagnostically marked with 7 yellowish and 8 black, strongly contrasting, longitudinal lines. The body is also marked with a series of thin dark-brown lines which run the length of the body. The head itself is hardly distinguishable from the body, being variegated black and brown, with a greyish-black chin. The venter is barred with black and greenish-yellow. The sub-caudals are red which are crossed by a single black bar. The scales are in 13 rows at the midbody. **Behaviour:** Nocturnal and terrestrial, living under rocks and vegetation. Feeds on snakes and lizards. **Range:** Present in peninsular monsoonal evergreen forest and Malayan mixed dipterocarp forest in peninsular Thailand up to 1,100 m. Recorded in Surat Thani, Nakhon Si Thammarat, Pattani, and Yala provinces. **Regional Range:** Thailand, Peninsular and Eastern Malaysia, Singapore, Brunei, Indonesia, and Philippines.

302

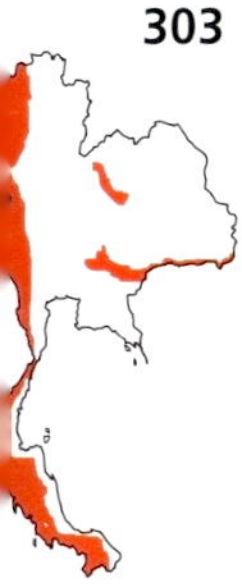

303 **Small-spotted Coral Snake** *Calliophis maculiceps*

Body Length: Up to 35.1 cm.

A small, slender, brick-red or pale brown snake with paired dark spots spaced regularly along the dorsal surface. The head is olive-brown. Black collar around the neck sometimes extending as a central vertebral stripe. The venter is reddish or pink. The tail has 2 black bands encircling it. The dorsal scales are in 13 rows. **Behaviour:** Terrestrial and nocturnal. Found near streams. Diet consists of other snakes, particularly blind snakes. **Habitat:** Inhabits a wide range of evergreen forest types. **Range:** Found throughout the country. **Regional Range:** Myanmar, Vietnam, Laos, Cambodia, Thailand, and Peninsular Malaysia.

303

Coral Snakes Genus Sinomicrurus

The coral snakes are elongate snakes. Their heads are indistinguishable from their bodies and they have short tails. They have small eyes with round pupils. They are all nocturnal. In Thailand 1 species is represented.

304 **McClelland's Coral Snake** *Sinomicrurus macclellandi*

Body Length: Up to 81.2 cm.

A medium-sized, cylindrical snake. The body is pinkish-red above with 23-40 narrow black bands around the body. The short, rounded head is black while the chin is cream-coloured. A broad, ivory-white band behind the eyes contrasts with a transverse black band immediately behind it. The venter is yellow with black marks. **Behaviour:** Terrestrial, preferring leaf litter. Nocturnal. It hunts other snakes and lizards. **Habitat:** Found in a variety of forest types up to 1,800 m. **Range:** Scattered records from northern, northeast (west) southeastern, and peninsular (northernmost) Thailand. Recorded in Doi Suthep National Park, Chiang Mai Province; Phu Luang Wildlife Sanctuary in Loei Province; Sakaerat near Khao Yai; Khao Ang Ru Nai Wildlie Sanctuary, Chachaengsao Province; and Kaeng Krachan National Park, Petchaburi Province. **Regional Range:** India, Nepal, Myanmar, China, Vietnam, Laos, and Thailand.

304

Sea Snakes Family Hydrophiidae

This family occurs in the tropical and sub-tropical regions of the Pacific and Indian Oceans. All members except one species living in a freshwater lake in the Philippines are marine. All species in Thailand are found in waters less than 200 m in depth. These snakes are entirely aquatic, and swim by having laterally compressed bodies. They have salt glands around the tongue which help to maintain osmotic balance. Most species eat adult fish. Thirteen species are found in Thailand.

KEY TO THE THAI HYDROPHIIDAE

1 a	At least 73 scale-rows around body	*Kolpophis*
b	Less than 73 scale-rows around body	**2**
2 a	Rostral shields broken up into 4-5 pieces, head shield with thickened edges	*Thalassophis*
b	Rostral shield single	**3**
3 a	Nasals separated by internasals	*Laticauda*
b	Nasals not separated by internasals	**4**
4 a	Ventrals much broader anteriorly than posteriorly	*Thalassophina*
b	Ventrals, if distinct, not much broader anteriorly than posteriorly	**5**
5 a	Elongate mental shield several times longer than broad	*Enhydrina*
b	Mental shield not much longer than broad	**6**
6 a	Head elongate; snout bill-like and flattened; gap of mouth very wide; normal colour pattern with yellow on ventral side and black on dorsal side	*Pelamis*
b	Colour pattern never with yellow on ventral side and black dorsal	**7**
7 a	Ventral (except on throat) divided into pairs of overlapping scales	*Disteira*
b	Ventrals usually entire or difficult to distinguish	**8**
8 a	Spines on rear edge of some head shields	*Acalyptophis*
b	No spines on rear edge of some head shields	**9**
9 a	Ventrals very small and difficult to distinguish; scales in lowermost scale-rows on flanks enlarged compared to dorsal scales	*Lapemis*
9 b	Ventrals distinct throughout and lowermost scale-rows on flanks not enlarged compared to dorsal scales	**10**
10a	Large ventrals, each at least three times as broad as adjacent body scales	*Aipysurus*
b	Small ventrals, each not more than two times as broad as adjacent body scales	**13**
11a	Less than 24 scale-rows around body	*Kerilia*
b	At least 24 scale-rows around body	**12**
12a	At least 14 maxillary teeth behind the poison fang	*Polydontognathus*
b	Less than 14 maxillary teeth behind the poison-fang	**13**

13a Interspace between bands narrower than bands posteriorly — *Leioselasma*

b Interspace between bands equal as bands — **14**

14a Anterior part of maxillare curved in the shape of an arc — *Hydrophis*

b Anterior part of maxillare not curved in the shape of an arc, tips of venomous fangs located below the level of maxillary teeth — *Chitulia*

Modified from Rasmussen (2001) and Kharin (2007).

Sea Kraits Genus *Laticauda*

This group has evolved from terrestrial elaphids. They are the only group of sea snakes that are oviparous and must return to land to breed. Their heads are barely distinguishable from the rest of the body. The nasals are separated by the internasals. The ventrals are large, being at least half as broad as the body. Worldwide: 3 species. Two species occur in Thai waters.

KEY TO THE THAI *LATICAUDA*

1 a Upper lip dark brown; no azygous shield between prefrontals; scale-rows number 19 around midbody — *L. laticaudata*

b Upper lip yellow; normally an azygous shield between prefrontals; scale-rows number 21-25 — *L. colubrina*

Key prepared by Chan-ard, Parr and Nabithabhata.

305 Yellow-lipped Sea Krait *Laticauda colubrina*

Body Length: Up to 360.0 cm.

A large, cylindrical-bodied sea snake, which is light or dark bluish-grey above with 24-64 black bands of uniform width along the entire length of the body: it shows considerable variation in neck and body coloration within its range. The head is black, which strongly contrasts with the upper lip, which is characteristically yellow or creamy; in the Black-banded Sea Krait the upper lip is dark. The crown also lacks any of the curved yellow markings which are found in the similar Black-banded Sea Krait. It is yellowish below, between the black bands. The tail is strongly compressed and vertically widened, and has 3-5 further black bands. The ventrals are large, being one-third the width of the body. The nostrils are lateral. It has 21-25 longitudinal midbody scale-rows; the Black-banded Sea Krait only has 19 rows. It is highly venomous, but is non-aggressive. **Behaviour:** Eats eels, particularly moray eels. Active during the day and night. During the breeding season, it congregates in large numbers on rocky islets. **Habitat:** Inhabits shallow seas including coral islands, coral reefs, mangroves, and open sea. **Range:** Currently only recorded from Phuket, peninsular

305

Thailand. **Regional Range:** India, Bangladesh, Myanmar, China, Japan, Vietnam, Thailand, Peninsular and Eastern Malaysia, Singapore, Brunei, Indonesia and Philippines, East Timor, Papua New Guinea, Australia, New Zealand, and the Pacific Islands.

---

306 Black-banded Sea Krait *Laticauda laticaudata*

Body Length: Up to 107.0 cm.

A large sea snake with a bluish-grey to ultramarine body above, with 20-70 black bands of uniform width throughout; these bands may or may not be complete below. The number of bands varies regionally. The black head has a curved yellow mark. The upper lip is characteristically dark, and not yellow. The venter is yellowish between the crossbands. It has only 19 scale-rows around the body. **Behaviour:** Active during the day and night. Preys on eels and fish. During the breeding season, congregates on islands. **Habitat:** Coral islands, coral reefs, mangroves, and the open sea. **Range:** Occurs in the

306

Gulf of Thailand along the coastline of lower peninsular Thailand. **Regional Range:** India, China, Japan, Thailand, Peninsular and Eastern Malaysia, Indonesia, Philippines, East Timor, Papua New Guinea, Australia, New Zealand, and Pacific Islands.

Sea Snakes Genus *Aipysurus*

Members of this genus have nostrils on the dorsal part of the head. The ventrals are large and distinct. Worldwide: 7 species. In Thailand, *Aipysurus eydouxii* is the sole representative.

307 **Spot-banded Sea Snake** *Aipysurus eydouxii*

Body Length: Up to 66.0 cm.

An intricately coloured sea snake, of almost uniform diameter throughout, and little compressed. The slender body is brownish or olive dorsally, with 39-55 pale yellow crossbands; these bands widen on the flanks. The adults have dark olive heads; the young have blackish heads. The venter is yellow. The ventral plates are large, like those of terrestrial snakes. The ventrals number 129-149. **Behaviour:** Swims slowly near the bottom. Feeds primarily on benthic fish eggs. **Habitat:** Inhabits shallow coastal waters, including estuaries and shallow bays. Likes turbid water between 30 and 50 m. **Range:** Gulf of Thailand. **Regional Range:** Vietnam, Thailand, Peninsular and Eastern Malaysia, Singapore, Indonesia, Papua New Guinea, New Caledonia, and Australia.

307

Sea Snakes Genus *Praescutata*

The head is short and distinct from the neck. The snout is broadly rounded. The ventrals are very broad anteriorly, and narrow posteriorly. This group is represented in Thailand by a single species, *Praescutata viperina*.

308 **Viperine Sea Snake** *Praescutata viperina*

Body Length: Up to 73.0 cm.

The head is short and distinct from the neck. The snout is broadly rounded. The body is sub-cylindrical with the greatest diameter in the posterior part of the body. Known in three colour forms. A common form is grey above and white below, with the two colours showing a clear line of demarcation where they meet. A second common phase is grey above with 25-35 black rhomboids usually more-or-less confluent or completely banded. A third rare form is completely banded. The ventral plates in the anterior portion of the body are much wider and larger than in the posterior part. The ventrals number 226-274. **Behaviour:** Diet consists of marine invertebrates. Ovoviviparous. **Habitat:** Marine. Found in coastal waters. **Range:** One of the commoner sea snakes in the Gulf of Thailand; also recorded in the Andaman Sea. **Regional Range:** Middle East, Pakistan, India, Sri Lanka, Bangladesh, Myanmar, China, Vietnam, Thailand, Peninsular and Eastern Malaysia, Singapore, and Indonesia.

308

Sea Snakes Genus *Lapemis*

Members of this group have a large head and stout body. The body scales are squarish or hexagonal, juxtaposed with the outer three rows larger than the others. The ventrals are small. In Thailand, the genus is represented by a single species.

309 Hardwicke's Sea Snake *Lapemis curtus*

Body Length: Up to 126.0 cm.

A short, rather stout, sea snake. The broad head is blackish with some yellow markings above the eyes. The chin and labials are blackish. The parietals are broken up. The body is yellowish with 35-55 dark olive bars across the back tapering to points on the sides; the interspaces are very narrow. There are seven bars on the tail. The lower part and end of the tail are black. The ventrals are very small and difficult to discern. The scale-rows on the sides of the body are enlarged compared to the dorsal scales. **Behaviour:** Active by day and night. Diet comprises eels, gobies, squid, and other invertebrates.

Ovoviviparous, with up to six neonates being produced. **Habitat:** Occurs in open sea, and mudflats, inhabiting localities with soft substrates of mud, or sand and mud. Recorded from waters from 4 to 25 m. May be found up rivers. **Range:** Occurs in the Gulf of Thailand and the Andaman Sea. One of the most abundant sea snakes. **Regional Range:** India, Sri Lanka, Bangladesh, Myanmar, China, Japan, Vietnam, Thailand, Peninsular and Eastern Malaysia, Singapore, Indonesia, Philippines, Papua New Guinea, New Caledonia, and Australia.

309

Sea Snakes Genus *Kolpophis*

This group is characterized by having a large head and stout body covered with small, irregular juxtaposed scales. A single species, *Kolpophis annandalei*, is represented in Thailand.

310 Annadale's Sea Snake *Kolpophis annandalei*

Body Length: Up to 91.0 cm.

A stout-bodied sea snake with a small head. The upper body is yellowish-olive with a series of 35-47 dark olive-grey bands. The dark bands are broader than the paler interspaces; these bands become less distinct with age. The dorsals are keeled. The head is variegated dark grey or dark olive, with lighter areas behind the eyes. The venter is creamish or whitish. The tip of the tail is olive-blackish. The scales are very small: there are always more than 73 scales around the middle of the body. The ventrals are distinct throughout, numbering 320-368. **Behaviour:** Diet consists of fish. **Habitat:**

310

Inhabits coasts, frequenting shallow waters and estuaries. **Range:** Found in coastal stretches in the Gulf of Thailand. Recorded in Songkhla Lake, in Songkhla Province. **Regional Range:** Vietnam, Cambodia, Thailand, Peninsular Malaysia, Singapore, Brunei, and Indonesia (Java).

Sea Snakes Genus *Disteira*

This group is represented by 2 species in Thailand.

KEY TO THE THAI *DISTEIRA*

1 a Ventrals not divided into pairs of overlapping scales; 1-2 maxillary teeth behind the poison fang — *D. nigrocincta*

b Ventrals (except the throat) divided into pairs of overlapping scales; 6-7 maxillary teeth behind the poison fang — *D. stokesii*

Key prepared by Chan-ard, Parr and Nabithabhata.

311 Black-banded Sea Snake *Disteira nigrocincta*

Body Length: Up to 108.0 cm.

Olivaceous to brownish above, with 40-60 narrow dark crossbands more or less of uniform width along the elongate, robust body; these are sometimes incomplete ventrally. The diameter of the neck is about one-third to one-half the greatest diameter of the body. The moderate-sized head is yellow with a dark streak along the upper lip. It has a dark triangular patch on the top of the head extending to the prefrontals. The eye is moderate. It has 1-2 maxillary teeth behind the poison fangs. The ventrals number 296-330. The dorsal scales number 39-45, which are imbricate throughout. **Behaviour:** Nothing is currently known of its natural history. **Habitat:** Considered to inhabit deep waters. **Range:** Found in the Andaman Sea, from Phuket to Krabi Province. **Regional Range:** India, Bangladesh, Myanmar, and Thailand.

312 Stoke's Sea Snake *Disteira stokesii*

Body Length: Up to 160.0 cm.

A large sea snake with a body girth reaching 25 cm. The enlarged head is almost uniformly brownish-black. The body is greyish-black with a series of 29 cream bands, which, except for the first, encircle the entire body. Each cream band has scattered black scales. The tail has five complete narrower cream bands and one incomplete band. The ventral plates are divided into two leaf-like scales. **Behaviour:** Eats benthic fish, foraging among coral. The species will apparently attack divers without provocation. Ovoviviparous. **Habitat:** Inhabits deep water, often associated with coral reefs. Found at depths from 4 to 60 m. **Range:** Occurs in the Gulf of Thailand and Andaman Sea. **Regional Range:** Pakistan, India, Sri Lanka, China, Vietnam, Thailand,

Peninsular and Eastern Malaysia, Singapore, Indonesia, Philippines, Papua New Guinea, New Caledonia, and Australia.

312

Sea Snakes Genus *Enhydrina*

This group is represented by two species, which have an elongate body and are among the largest sea snakes in the world. The ventrals are small, and distinct throughout.

313 **Beaked Sea Snake** *Enhydrina schistosa*

Body Length: Up to 140.0 cm.

One of the largest sea snakes, with a beak-like rostral. The top of the head, the dorsal surface of the elongate body, and the tail are dark slate-grey, sometimes with indistinct darker marking forming 40-60 transverse bands. The sides are creamy yellow. The venter is cream to white. The posterior end is strongly compressed. The young are grey or bluish-grey above, whitish below, with dark grey or black annuli. One of the most poisonous

313

snakes in the world. **Behaviour:** Active at night. Cruises along near the sea bottom, eating shrimps, catfish, and other fish. Detects prey through pressure waves rather than vision. **Habitat:** Occurs in shallow open sea, river mouths, estuaries, and mangroves. Known to move long distances upstream. Occurs in water depths from 3.7 to 22 m. Likes soft, muddy, or sandy substrates. **Range:** Found in the Gulf of Thailand and the Andaman Sea. It is the most common sea snake in the Chao Phraya estuary in the dry season. **Regional Range:** Middle East, Pakistan, India, Sri Lanka, Bangladesh, Myanmar, Vietnam, Cambodia, Thailand, Peninsular and Eastern Malaysia, Singapore, Indonesia, Philippines, Papua New Guinea, and Australia.

Sea Snakes Genus *Kerilia*

The sole representative of this genus, *Kerilia jerdonii,* has a robust body and a small head. The scales on the body are keeled and in regular rows. The poison fang is moderate, followed by a short or absent interval and 7-9 teeth.

314 **Jerdon's Snake** *Kerilia jerdonii*

Body Length: Up to 96.0 cm.

The short head is grey or greenish above, yellowish or white below. The body is almost of equal diameter throughout, being olive, blue-grey to ultramarine dorsally, with 30-38 wide black bands, which narrow laterally and may encircle the entire body. The sides and the ventral surface are yellowish. The tail is marked similarly to the rest of the body, with a black tip always. The diastema between the fangs and the maxillary teeth is either absent or very short. It has fewer than 23 scale-rows around the body; most other sea snakes have more than this. **Behaviour:** It is a crevice forager, swimming slowly near the bottom, periodically poking its head into

314

crevices and holes. **Habitat:** Found near coral reefs. **Range:** Found in the Gulf of Thailand and the Andaman Sea. **Regional Range:** India, Sri Lanka, Bangladesh, Myanmar, China, Vietnam, Thailand, Peninsular and Eastern Malaysia, Singapore, and Indonesia (Sumatra).

Sea Snakes Genus *Thalassophis*

This is a monotypic genus, represented by *Thalassophis anomalus*. The nostrils are supra-lateral and the nasals are separated by a pair of elongated internasals.

315 **Anomalous Sea Snake** *Thalassophis anomalus*

Body Length: Up to 81.0 cm.

A moderate-sized sea snake with a short head. The stout body is pale grey above, with 30-36 dark crossbars that are much broader than their interspaces, which either taper to points on the sides or continue as narrow bands across the venter. The ventral surface is white. The young of this species has a pale mark across the snout extending backward on either side of the head. **Behaviour:** Little known. **Habitat:** Found in estuaries. **Range:** Eastern parts of the Gulf of Thailand. **Regional Range:** Vietnam, Thailand, Peninsular and Eastern Malaysia, Singapore, and Indonesia.

Sea Snakes Genus *Acalyptophis*

This genus has sub-imbricate scales on the body, in 23-29 rows. The ventrals are small, but distinct. It is represented by one species, *Acalyptophis peronii*.

316 **Peron's Sea Snake** *Acalyptophis peronii*

Body Length: Up to 86.1 cm.

A thick-bodied sea snake, with a short, rounded pale brown head. The plates around the eyes bear characteristic spines. It is light brown or grey above with a series of transverse, greyish-black blotches, which narrow down the sides and the venter. The tail is strongly compressed and oar-shaped. It has 19-23 scale-rows on the neck, and 23-29 scale-rows on the thickest part of the body. The ventrals number 142-206. **Behaviour:** Swims slowly near the bottom, feeding on gobies and other fish. **Habitat:** Sandy areas among coral reefs. **Range:** Found in the Gulf of Thailand. **Regional Range:** China, Vietnam, Thailand, Indonesia (Salawati and Papua), Papua New Guinea, New Caledonia, and Australia.

316

Sea Snakes Genus *Pelamis*

These sea snakes have short, stoutish bodies, covered with hexagonal or squarish, juxtaposed scales. There are 49-67 scales around the thickest part of the body. The ventrals are very narrow, being divided by a longitudinal grove. The group is represented by a single species, which is the only pelagic sea snake in the world.

---

317 Yellow-bellied Sea Snake *Pelamis platura*

Body Length: Up to 113.0 cm.

Colour and patterns vary widely. The head is strongly elongated and flattened, being black above, brown or brownish below. The snout looks like a duck's bill. The gape opening is very wide. The body has a longitudinal black band on the dorsum, the outer adjacent rows are yellow forming a distinct colourful stripe; the rest of the body is brown. The tail is yellow with dorsal and ventral black bars. Ventral plates are absent. **Behaviour:** A pelagic sea snake, using an ambush strategy to capture fish. Often found in association with floating debris. Often lies motionless on the surface. Can swim backward as well as forward. **Habitat:** Most pelagic of the sea snakes, occurring in the open sea well away from the coasts and reefs. It has occasionally been recorded in intertidal habitats, including mangrove. **Range:** Found in the Gulf of Thailand and the Andaman Sea. **Regional Range:** Africa to Mexico, all tropical waters.

317

Sea Snakes Genus *Hydrophis*

These sea snakes have long bodies with imbricate, sub-imbricate, or juxtaposed scales, with 29-57 dorsal rows at the thickest part of the body. The head shields are entire. The poison fangs are followed after an interval by 1-18 maxillary teeth. The ventrals are usually distinct throughout. This group is represented by eight species in Thailand.

KEY TO THE THAI *HYDROPHIS*

1	**a**	Ventrals are divided by the elongated groove in the posterior part of the trunk. The maxillare protrudes before the palatinum. The parietale is connected with the praefrontale and separates the frontale from the orbital edge and the postorbitale. There are no slit-like scales on the infralabials	**2**
	b	Ventrals are smooth in the posterior part of the trunk. The maxillare does not protrude before the palatinum. The parietale is connected with the praefrontale either in whole and separates the frontale from the edge of the orbit and from the postorbotale or connected incompletely and then the frontale contacts partially with the orbital margin and with the postorbitale. Slit-like scales are usually present	**3**
2	**a**	Prefrontal plates contact the 3rd upper labial. 41-48 scales around the midbody. 404-468 ventrals	*H. cantoris*
	b	Prefrontals plates do not contact the 3rd upper labial. 30-43 scales around the midbody. 220-350 ventrals	*H. gracilis*
3	**a**	The parietale always connected with the praefrontale so that the frontale separated from the orbitale edge and the post-orbital	**4**
	b	The parietale is not always completely connected with the praefrontale so that the frontale enters partially into the margin of the orbit and contacts the post-orbital	**5**
4	**a**	Slit-like scales are absent on infralabials. Body scales smooth or with little keel. 360-430 ventrals. The pterygoideum with 7-10 teeth, 12-13 teeth on the dentale	*H. klossi*
	b	Slit-like scales are usually present on infralabials. Body scales with strong keel 266-368 ventrals. The pterygoideum with 9-12 teeth, 13-14 teeth on the dentale	*H. melanosoma*
5	**a**	There are 19-23 scales around the neck. The palatinum always with prominence for the maxillare. 34-60 crossbands around the body and tail	*H. obscurus*
	b	There are more than 23 scales around the neck; if they number 23, the palatinum is without prominence for the maxillare. Ventrals always more than 320	**6**

6 a The head dark with yellow strip through the snout and along its sides. 37-45 scales around the midbody. 328-453 ventrals — *H. brookii*

b The head black or dark olive with yellowish oval specks on each side. 39-58 scales around the midbody. 323-514 ventrals — **7**

7 a 27-33 scales around the neck and 49-58 scales around the midbody. 410 -514 ventrals (usually more than 455) — *H. fasciatus*

b 25-30 scales around the neck and 39-49 scales around the midbody. 323-453 ventrals (usually less than 366) — *H. atriceps*

Key modified from Taylor (1965); Tweedie (1983).

318 Gunther's Sea Snake *Hydrophis cantoris*

Body Length: Up to 174.0 cm.

A large, dark brown or grey sea snake with light brown or yellowish crossbands or lateral spots. The head is brown, grey, or yellowish-green; in young individuals it is black. Distinguished from most other members of the genus *Hydrophis*—apart from the Graceful Sea Snake—by having the ventrals divided by an elongated groove in the posterior part of the trunk. It also lacks slit-like scales on the infralabials. This species has 41-48 scales around the midbody; by comparison the Graceful Sea Snake has 30-43 scales around the midbody. It has 404-468 ventrals; the Graceful Sea Snake only has 220-350 ventrals. **Behaviour:** Diet includes eels and invertebrates. **Habitat:** Marine. **Range:** Andaman Sea. **Regional Range:** Pakistan, India, Sri Lanka, Bangladesh, Myanmar, Thailand, and Peninsular Malaysia.

319 Graceful Sea Snake *Hydrophis gracilis*

Body Length: Up to 93.0 cm.

A very small-headed sea snake, with a slender anterior portion. The diameter of the body is 3-5 times that of the neck. The body is bluish-greyish above and pale yellow below, with 40-60 dark crossbands and spots. The head is olive or yellowish. Distinguished from most other members of the genus *Hydrophis*—apart from Gunther's Sea Snake—by having an elongated groove dividing the ventrals along the posterior part of the body. It also lacks slit-like scales on the infralabials. The Graceful Sea Snake has 30-43 scales around the midbody; by comparison the Gunther's Sea Snake has 41-48 scales around the midbody. It has 404-468 ventrals; the Gunther's Sea Snake only has 220-350 ventrals. **Behaviour:** Eats fish, mainly eels. Forages in sand, probing with its head. **Habitat:** Mangrove and around coastal reefs. Likes deeper water between reefs. **Range:** Gulf of Thailand and Andaman Sea. **Regional Range:** Middle East, Pakistan, India, Sri Lanka, Bangladesh, Myanmar, China, Thailand, Peninsular and Eastern Malaysia, Singapore, Indonesia, Papua New Guinea, and Australia.

319

320 Kloss's Sea Snake *Hydrophis klossi*

Body Length: Up to 119.0 cm.

A robust snake with a small dark head, and a narrow, anterior body. The head is usually characteristically marked with a pronounced yellow horseshoe. The body has 50-75 black bands with green or grey narrower interspaces. The venter is yellowish. It has 360-430 ventrals, which are smooth in the posterior part of the trunk; lacking an elongated groove. There are 23-27 scales around the neck, and 37-45 scales around the midbody. **Behaviour:** A crevice feeder. **Habitat:** Entirely marine, favouring deep water. **Range:** Gulf of Thailand and Andaman Sea. **Regional Range:** Thailand, Peninsular Malaysia, Singapore, and Indonesia (Sumatra).

320

321 Black Sea Snake *Hydrophis melanosoma*

Body Length: Up to 139.0 cm.

Greyish above, with 50-70 wide crossbands. The head and neck are black, with narrow, almost oval, yellowish markings passing through the nasals, prefrontals, preoculars, postoculars, and temporals; sometimes these markings join on the occiput. It has 266-368 ventrals; these lack an elongated groove in the posterior part of the trunk. The body scales have strong keels. The pterygoideum with 9-12 teeth. **Behaviour:** Eats fish, mainly eels. **Habitat:** Shoal waters. Likes deep water. **Range:** Occurs in the Gulf of Thailand and the Andaman Sea. **Regional Range:** Thailand, Peninsular and Eastern Malaysia, Singapore, Brunei, Indonesia, Papua New Guinea, and Australia.

321

322 Daudin's Sea Snake *Hydrophis obscurus*

Body Length: Up to 109.0 cm.

Greyish or whitish, with 30-60 black or brown crossbands which are much wider than the greyish ground colour along the entire length of the body and tail. In adults these markings become indistinct and older individuals are almost uniform greyish above. The body scales are smooth. The head is black or dark brown, with a yellow horseshoe marking extending from the snout to the occipitals. The ventrals number 298-346. **Behaviour:** Little is known of the behaviour. **Habitat:** Marine. **Range:** Found throughout the Andaman Sea. **Regional Range:** India, Sri Lanka, Bangladesh, Myanmar, and Thailand.

323 Brook's Sea Snake *Hydrophis brookii*

Body Length: Up to 96.5 cm.

A medium-sized sea snake. It has a very small black head with a yellow band across the snout. The neck and the anterior portion of the body are also small, while the broader, posterior portion of body is compressed laterally. Body coloration is greyish above and yellowish-white below. It has 60-80 blackish bands along the length of the body and tail. The anterior bands are twice as broad as the interspaces, are of even width, and completely encircle the body; the posterior bands narrow on the sides. There are 37-45 scale-rows at the middle of the body. **Behaviour:** A crevice forager. **Habitat:** Entirely marine. **Range:** Gulf of Thailand and Andaman Sea. **Regional Range:** Vietnam, Thailand, Peninsular and Eastern Malaysia, Singapore, and Indonesia.

323

324 Striped Sea Snake *Hydrophis fasciatus*

Body Length: Up to 101.0 cm.

A sea snake with a very variable body coloration and patterning. It has a very small black head and chin, with a narrow black stripe along the neck and the particularly narrow anterior part of the body. It has 56 olive-black bands with greenish-white or cream interspaces; these bands narrow laterally becoming obsolete on the belly. It has more than 410 ventral scales. **Behaviour:** It is a crevice forager. **Habitat:** Entirely marine. Inhabits deep water. **Range:** Gulf of Thailand and Andaman Sea. **Regional Range:** Pakistan, India, Sri Lanka, Bangladesh, Myanmar, Thailand, Peninsular Malaysia, Singapore, and Indonesia (Sumatra).

324

Sea Snakes Genus *Leioselasma*

This group has distinctly imbricate scales throughout the large body, with more-or-less rounded tips posteriorly. They have 5-6 maxillary teeth behind the poison fangs. The two infralabials are unsegmented. The scales number 37-47 at mid-body. It is represented by one species in Thailand.

325 Blue-banded Sea Snake *Leioselasma cyanocinctus*

Body Length: Up to 175.0 cm.

A large sea snake. The body is greyish to dusky-olive dorsally, pale yellow ventrally, with 48-65 black rings which usually broaden on the back and belly. Occasionally a few rings may be forked on one side of the body. The tail has 5-8 rings and is black on the tip. **Behaviour:** Feeds upon marine fishes. **Habitat:** Found in shallow, muddy mangrove during the wet season but may also be found in deep water. **Range:** Found in the Gulf of Thailand and the Andaman Sea. **Regional Range:** Middle East, Pakistan, India, Sri Lanka,

325

Bangladesh, Myanmar, China, Japan, Vietnam, Thailand, Peninsular and Eastern Malaysia, Singapore, Indonesia, and the Philippines.

Sea Snakes Genus *Chitulia*

This group is represented by 6 species in Thailand.

KEY TO THE THAI *CHITULIA*

1 a	The neck and first one-third of the body are moderately or very narrow, and are cylindrical in shape. Head is dark-grey or olive-green. There are 16-19 teeth on the dentate. There are 242-343 ventral shields	*C. torquata*
b	The neck and first one-third of the body are wide and rounded in transverse section	**2**
2 a	There is a series of slit-like scales on the sublabial shields, from behind the second shield backward. There are 288-404 ventrals	*C. lapemoides*
b	Usually slit-like scales on the sublabial shields are absent. If a single slit-like scale is present, it is located between two labial shields. There are 195-336 ventral shields	**3**
3 a	There are 7-8 maxillary teeth. There are 29 or fewer scales around the neck. Scales have a keel consisting of two tubercles located one behind the other	*C. bituberculata*
b	There are 9-13 maxillary teeth. There are 29 or more scales around the neck	**4**
4 a	There are 17 teeth on the dendate. There are 163-168 trunk vertebrae. The width of the head between the eyes is less than half the head length from snout tip to the edge of the parietal shields. Adult individuals lack dorsal spots which may be present in young animals	*C. inornata*
b	There are 18-23 teeth on the dentale. The width of the head between the eyes is equal to or larger than one-half of the head length from snout tip to the edge of the parietal shields. Adult animals have dark dorsal spots or stripes	**5**
5 a	Anterior half of the body bears transverse stripes that are wider than the gaps between them on the back (by two scales or less) and narrrower than the latter along the lateral sides. There are 153-163 trunk vertebrae	*C. ornata*
b	The stripes are represented by rounded spots until the middle of body. The distance between the spots on the back is more than two scales. There are 164-173 trunk vertebrae	*C. lamberti*

Key modified from Kharin (2007).

326 Black-headed Sea Snake *Chitulia torquata*

Body Length: Up to 104.5 cm.

The elongate body is compressed posteriorly. It is greyish or greenish-grey above, with 55-68 dark grey or black annuli which may be incomplete ventrally. Whitish yellow below. The moderate-sized head is black to dark olive with yellow markings across the snout and the sides of the head. It sometimes has yellow spots on the frontal and parietal shields. It has 271-343 ventrals. The young are whitish, with 46-63 well-defined black bands. **Behaviour:** It is a crevice forager. **Habitat:** Inhabits brackish waters. **Range:** Occurs in the Gulf of Thailand. **Regional Range:** Vietnam, Cambodia, Thailand, Peninsular and Eastern Malaysia, Singapore, and Indonesia.

326

327 Persian Gulf Sea Snake *Chitulia lapemoides*

Body Length: Up to 87.0 cm.

The head is moderate, being greenish-brown with or without a yellow curved mark above. The body is neither particularly elongated nor compressed, being whitish with strongly dilated, blackish-grey bands dorsally which disappear ventrally. The tail is also banded with a black tip. In some individuals the bands are connected or even fused. The juveniles are completely banded. **Behaviour:** Swims along sea bottom, preying on fish. **Habitat:** Marine. Found in deep water with gravel bottoms at depths of 30 m. **Range:** Gulf of Thailand and Andaman Sea. **Regional Range:** Middle East, Pakistan, India, Sri Lanka, Bangladesh, Thailand, Peninsular Malaysia, and Singapore.

327

328 Biturberculed Sea Snake *Chitulia bituberculata*

Body Length: Up to 112.0 cm.

A large member of the genus with a small head and elongate body being twice the diameter of the neck. The body, which is markedly compressed posteriorly, is uniformly grey above, yellowish below. It has 48 scale-rows on the body. The ventrals number 276. **Behaviour:** It is a crevice forager. Diet includes eels. **Habitat:** Entirely marine. **Range:** Found in the Andaman Sea. **Regional Range:** Sri Lanka and Thailand.

329 Inornate Sea Snake *Chitulia inornata*

Body Length: Up to 70.0 cm.

The adult is grey above, and white below, with the two colours meeting on the flanks in an irregular line of demarcation. The head is narrower than the Reef Sea Snake, the breadth between the eyes being less than half the length to the end of the parietals. It has 195-278 ventrals. It has 10-12 maxillary teeth behind the fangs. **Behaviour:** Cruises along the sea bottom. Eats fish. **Habitat:** Sometimes ascends rivers. **Range:** Occurs in the Gulf of Thailand and the Andaman Sea. **Regional Range:** Thailand, Indonesia (Java) and the Philippines.

330 Reef Sea Snake *Chitulia ornata*

Body Length: Up to 70.0 cm.

The head is large and robust, being olive-green or greyish above and whitish below. The body is also olive-green above, whitish below. The body has 30-56 broad, dark greyish bands with characteristic, very narrow light-grey interspaces, which are almost equidistant anteriorly. The tail has 6-11 dark greyish bands with narrow white interspaces. **Behaviour:** Cruises along the sea bottom. Eats fish. **Habitat:** Marine. Inhabits coral reefs. **Range:** Gulf

330

of Thailand and Andaman Sea. **Regional Range:** Middle East, Pakistan, India, Sri Lanka, Bangladesh, Myanmar, China, Japan, Vietnam, Thailand, Peninsular and Eastern Malaysia, Singapore, Indonesia, Philippines, Papua New Guinea, Australia, and Pacific Islands.

---

331 Lambert's Sea Snake *Chitulia lamberti*

Body Length: Up to 95 cm.

The anterior dorsal part of the body bears rounded spots. The width between the spots is greater than the width of two scales. There are 42 or more transverse stripes on the body and tail. By comparison, the Reef Sea Snake has fewer than 42 stripes on the body and tail. **Behaviour:** Unknown. **Habitat:** Entirely marine. Found in deep water. **Range:** Gulf of Thailand and Andaman Sea. **Regional Range:** Thailand, Singapore, and the Philippines (Gigante).

Sea Snakes Genus *Polydontognathus*

The genus is monotypic, being represented by *Polydontognathus caerulescens* which is found in Thailand. It has 14-18 maxillary teeth following the poison fangs.

---

332 Dark Blue-banded Sea Snake *Polydontognathus caerulescens*

Body Length: Up to 88.3 cm.

A medium-sized sea snake. It has a moderately small black or dark brown head, sometimes with a yellow horseshoe-shaped mark on top. The body is bluish-grey or greyish dorsally and yellowish-white below. It has 39-63 bluish-grey bands on the body and the tail. On the back these stripes are wider than the gaps between the stripes; on the lateral sides they are narrower. Sometimes the transverse stripes are absent, being replaced by triangular or rounded spots. It has 13-18 maxillary teeth. **Behaviour:** Entirely marine. Cruises along the sea bottom, preying mainly on gobies, eels, and other fish. **Habitat:** Appears to like muddy bottoms. Found in waters as shallow as 3.7 m. **Range:** Gulf of Thailand and Andaman Sea. **Regional Range:** Pakistan, India, Sri Lanka, Bangladesh, Myanmar, China, Vietnam, Thailand, Peninsular and Eastern Malaysia, Singapore, Indonesia, and Australia.

Vipers Family Viperidae

This family comprises the vipers. They have stout bodies. The pupil is vertical. They are venomous, possessing a venom gland behind the eye. All species in this family have a pair of long, curved, hinged fangs located on the upper jaw at the front of the mouth. All are terrestrial. Thirteen species are found in Thailand.

Vipers Genus *Daboia*

The vipers are characterized by having heads distinct from the neck, and having moderately sized bodies. The snout is short and rounded, with a protruding upper edge. The head scales, particularly the parietals and the frontal, are small and irregular. The eye is moderate with vertical pupils. The group is represented in Thailand by a single species, the Siamese Viper.

333 Siamese Viper *Daboia siamensis*

Body Length: Up to 127.0 cm.

A large snake reaching more than a metre in length. Light brown to buff-brown body with three longitudinal rows of brown oval spots bordered with black and edged with light buff-white. There are also small irregular black spots along the flanks. The head is bluntly rounded and distinct from the body. The venter is uniformly yellowish-white with black spots. **Behaviour:** Crepusclar and nocturnal. Ground-dwelling. Diet consists of rodents, birds, lizards, and frogs. **Habitat:** Occurs in open dry habitats and grasslands in lowlands. **Range:** Recorded in all provinces in central Thailand. **Regional Range:** Myanmar, China, Vietnam, Cambodia, Thailand, and Indonesia (Java).

333

Pit Vipers Family Crotalidae

This family comprises the Malayan and other Pit Vipers. The majority have a very distinct head. They are sometimes terrestrial, though many of them are arboreal, preferring the understorey as they await prey. The family is subject to intense taxonomic scrutiny and regular revision of species. Twelve species are currently recognized in Thailand.

KEY TO THE THAI CROTALIDAE

1	**a**	9 normal symmetrical head shields	*Calloselasma rhodostoma*
	b	Upper part of head covered for the most part with small scales except for presence in some species of supraoculars	**2**
2	**a**	12–14 strongly keeled scales between small supraoculars; 9-10 supralabials, 1st free from nasal scale, 2nd not bordering labial pit	*Tropidalaemus wagleri*
	b	9-15 obtusely keeled or smooth scales between supraoculars; 2nd labial bordered loreal pit; 1st fusing with nasal or separated from nasal	**3**
3	**a**	1st supralabial completely separated from nasal	**4**
	b	1st supralabial partly or completely joined to nasal	**13**
4	**a**	Head scales flat, partly imbricate, larger; 5-8 scales (rarely 9) between large supraoculars	**5**
	b	Head scales smaller, subimbricate, 9-13 scales between narrow supraoculars	**7**
5	**a**	Head short; tail not prehensile; 7-10 supralabials; 2-4 rows of scales between supralabials and suboculars. Brown with large spots of darker brown; scale-rows at mid-body 23-25 (very rarely 27); 132-158 ventrals	*Ovophis convictus*
	b	Head elongate; tail somewhat prehensile; 1-2 rows of scales between labials and sub-ocular (3rd labial may touch sub-ocular); scale-rows at mid-body 21. Green larger snakes	**6**
6	**a**	Body marked with well-defined dark crossbands in adults, nearly totally green in juveniles; dorsal head surface heavily marked with black; three supralabials (3-4-5) in contact with sub-ocular, supranasals usually separated; 186-193 ventrals; 55-71 sub-caudals	*Parias sumatranus*
	b	Body marked with two rows of rounded whitish or pinkish spots; upper head surface not marked with black; one or two supralabials (3 or 3-4) in contact with a subocular; supranasals in contact; 177-198 ventrals; 63-85 subcaudals	*P. hageni*
7	**a**	General colour green with a lateral light stripe on outer scale-row bordered above by an orange or chocolate-brown strpe	**8**
	b	General color brownish-grey with a dorsal series or irregular brown spots, smaller spots on sides; whitish below; scale-rows 19,19,15	**12**

8 a	Hemipenis strongly spinosed	**9**
b	Hemipenis without spines	**10**
9 a	Tail tip with red blotches	*Viridovipera gumprechti*
b	Tail tip with grey or white colour	*V. vogeli*
10a	General colour green in life	**11**
b	Not as above	**15**
11a	Tail with reddish stripe	**12**
b	Tail with reddish blotches	**13**
12a	The head and dorsal surface are green, the ventral surface is greenish- or yellowish-white; the supralabials are pale green, yellow, or white, as are the chin and the throat	*Cryptelytrops albolabris*
b	Top of the head and dorsal surface of the body are bluish-green; labials are light bluish-green, chin and throat are bluish-white	*C. macrops*
13a	Postocular streak present	**14**
b	Postocular streak absent	*Popia nebularis*
14a	Eyes deep red (in male); usually no crossbands on the body, only rarely with white vertebral dots	*P. popeiorum*
b	Eyes greenish-yellow, gold, or yellow copper (in male); dark bands on the body, usually white vertebral dots present	*P. fucatus*
15a	Body pattern with series of irregular crossbars	**16**
b	Body pattern with series of spots or blotches	**17**
16a	Two pairs of small internasals; found only in western Thailand, Kanchanaburi Province	*Cryptelytrops kanburiensis*
b	One pair of elongated internasals; found in upper part of peninsular in Surat Thani and Nakhon Si Thammarat provinces	*C. venustus*
17a	Purplish-brown uniform, olivaceous or greyish above, variegated or spotted and marked with brown	*C. purpureomaculatus*
b	Olive-brown, dark greyish-brown, or dark yellowish-grey, with about 22-35 large, dorsolateral dark-brown, dark reddish-brown, dark purple-brown, or dark greyish-brown blotches	*Trimeresurus wiroti*

Key modified from Taylor (1965); Tweedie (1983); David & Vogel (1996).

Pit Vipers Genus *Calloselasma*

Members of this group have nine symmetrical head shields. The eye has a vertical pupil. It has a lateral sensory pit on the snout. The maxillary has a very large fang but no teeth. The body is nearly cylindrical, with smooth or keeled scales. The tail is moderate or short. It is represented by a single species, the Malayan Pit Viper.

334 Malayan Pit Viper *Calloselasma rhodostoma*

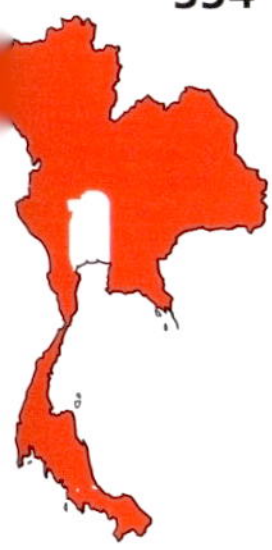

Body Length: Up to 70.0 cm.

The large, strongly triangular head is grey-brown with a light, dark-bordered stripe down each side. The stout body is also greyish-brown with a series of paired, dark brown, spots on both sides with a narrow blackish line bordered by a bluish-white line. The venter is brownish-white. The snout is pointed and slightly upturned. The maxillary comprises a very large fang and no other teeth. The eye has a vertical pupil. **Behaviour:** Nocturnal. Ground-dwelling, living in the leaf litter. Generally slow-moving, relying on its camouflage to take prey. **Habitat:** Found in evergreen forest, south-eastern and peninsular monsoonal evergreen forest, and Malayan mixed dipterocarp forest. **Range:** Widespread throughout the country, at lower altitudes. **Regional Range:** Vietnam, Laos, Cambodia, Thailand, Peninsular Malaysia, and Indonesia (Java).

334

Mountain Pit Vipers Genus *Ovophis*

Members of this group are small to medium-sized, ground-dwelling snakes found at higher elevations. They are characterized by having fewer than 35 subcaudals which are divided or entire. They have stout bodies with triangular heads. The Mountain Pit Viper is the only representative of the genus in Thailand.

335 Mountain Pit Viper *Ovophis convictus*

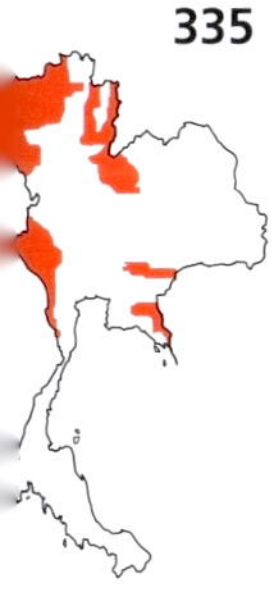

Body Length: Up to 110.0 cm.

A short-bodied viper reaching just over a metre in length. The body is brown with 22 large darker-brown spots. The top and sides of the short head are blackish-brown. It has a brown stripe behind the eye. Venter is dull yellow. The tail is light brown, becoming greenish-yellow toward the end, and is not prehensile. Scale-rows at mid-body number 25-27. **Behaviour:** Terrestrial.

Habitat: Inhabits upper montane and lower montane forests. **Range:** Occurs in mountains of northern, northeastern (western), southwestern, and southeastern Thailand. **Regional Range:** Vietnam, Laos, Cambodia, Thailand, Peninsular Malaysia, Singapore, and Indonesia (Sumatra).

335

Green Pit Vipers Genus *Cryptelytrops*

Members of this group are small to medium-sized snakes, all of which have heads very distinct from the neck. The eye has a vertical pupil. The chin-shields are smooth. The tail may be short or moderately long, and prehensile. The sub-caudals are usually divided. They are crepuscular or nocturnal, with arboreal habits. Most species are ovoviviparous. Five species occur in Thailand.

---

336 White-lipped Green Pit Viper *Cryptelytrops albolabris*

Body Length: Up to 87.4 cm.

Body is green with numerous irregular transverse dark spots or bands, each more or less reaching the venter. It has a light lateral whitish stripe, which is weak or absent in females. The prehensile tail is mottled brown, and not red or spotted. Head is brownish. **Behaviour:** Nocturnal and arboreal. Diet consists of rodents, birds, lizards, and frogs. **Habitat:** Often found close to streams. Found in all forest types, although presence in Malayan mixed dipterocarp forest unconfirmed, up to at least 1,000 m. **Range:** Found throughout the country, apart from central plains. **Regional Range:** Myanmar, China, Vietnam, Laos, Cambodia, Thailand, Singapore, and Indonesia.

336

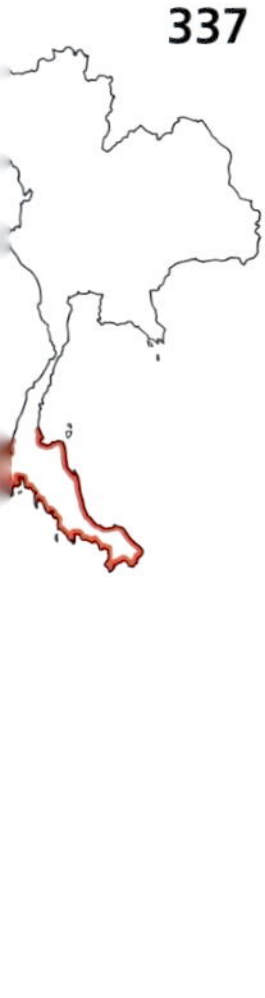

337 Mangrove Pit Viper *Cryptelytrops purpureomaculatus*
Body Length: Up to 90.0 cm.

Attractively coloured viper with dark purplish-brown on pale green. The top of the head is blackish-brown. Tail is prehensile. **Behaviour:** Usually arboreal, found in low trees. Diet includes small mammals. **Habitat:** Occurs in mangrove. **Range:** Found in coastal areas throughout peninsular Thailand. **Regional Range:** Myanmar, Thailand, Peninsular Malaysia, Singapore, and Indonesia (Sumatra).

337

338 Large-eyed Pit-Viper *Cryptelytrops macrops*

Body Length: Up to 71.0 cm.

Distinguished from all other pit vipers by having a bluish-green body and bluish-green top to the head. The labials are light bluish-green. It also has striking clean white lines along the entire length of the lower sides. The terminal part of the tail is reddish-brown. **Behaviour:** Crepuscular and nocturnal. Arboreal. Diet consists of rodents, birds, lizards, and frogs. **Habitat:** Occurs in dry evergreen forest and southeastern monsoonal evergreen forest. **Range:** Found in central and southeastern Thailand. **Regional Range:** Vietnam, Cambodia, and Thailand.

338

339 Kanburi Pit-Viper *Cryptelytrops kanburienis*

Body Length: Up to 66.7 cm.

Sexually dimorphic. The moderately elongated body of the male is drab olive-brown or greyish-green with vertebral white dots every 3-5 scales. The stout-bodied female is light greyish-brown with irregular zigzag-like dark brown or dark olive-brown crossbands. The head, which is rather short and ovoid, has the same coloration as the body, marbled with a net of blotches of the same colour as the dorsal crossbands. The eyes are brown, with a black vertical pupil. It has a moderate canthus rostralis. The infralabials, chin, and throat are whitish-brown. The venter is cream below. The tail is prehensile. **Behaviour:** Diurnal and terrestrial. Diet unstudied. **Habitat:** Found in limestone hill forest. **Range:** Occurs in Sai Yok National Park, Kanchanaburi Province, western Thailand. **Regional Range:** Endemic to Thailand.

339

340 Brown-spotted Pit Viper *Cryptelytrops venustus*
Body Length: Up to 68.0 cm.

The dorsal and lateral body in both sexes is dark green, olive-green, or bottle-green with irregular, zigzag-like, reddish-brown or purplish-red crossbands. The head and temporal regions have a similar coloration to the body. The head has a wide, conspicuous dark purplish-brown postocular streak. The eyes are yellowish-brown or gold. The infralabials, chin, and throat are greenish-white or pale green, with some rusty-brown blotches. The belly is pale green; the ventral surface of the tail becomes reddish-brown posteriorly. **Behaviour:** Diurnal and terrestrial. Diet unknown. **Habitat:**Occurs in peninsular monsoonal evergreen forest. **Range:** Recorded Surat Thani and Nakhon Si Thammarat provinces in peninsular Thailand. **Regional Range:** Endemic to Thailand.

340

Green Pit Vipers Genus *Viridovipera*

Members of this group are small to medium-sized snakes, and all have heads that are very distinct from the neck. The eye has a vertical pupil. The chinshields are smooth. The tail may be short or moderately long and prehensile. The sub-caudals are usually divided. They are crepuscular or nocturnal, with arboreal habits. Most species are ovoviviparous. Two species occur in Thailand.

341 **Gumprecht's Green Pit Viper** *Viridovipera gumprechti*

Body Length: Up to 120.0 cm.

A large pit viper. Overall bright-green coloration in males, deep or dark green in females. It has a conspicuous bi-coloured postocular stripe in males, which is thin and white below, wide and red above; in females this stripe is only white. It also has a vivid, bi-coloured ventrolateral stripe running under the first third of the tail, bright or deep red below and white above in males; white or blue but well-defined in females. The interstitial skin is black. The greatest part of the dorsal of the tail is rusty-red. The eyes are deep red in males; yellow in females. **Behaviour:** Noctural and arboreal. Diet consists of rodents and skinks. Often found in bamboo thickets near streams. **Habitat:** Found in sub-montane forest, mainly between 800 and 1200 m. **Range:** Found in northeastern Thailand, in Loei, Phitsanulok, Petchabun and Chaiyaphum provinces. **Regional Range:** Laos and Thailand.

341

342 **Vogel's Pit Viper** *Viridovipera vogeli*

Body Length: Up to 130.0 cm.

One of the largest Asian pit vipers, with the female reaching 130 cm and the male reaching 95 cm. The male has a dark-green dorsum with faint dark crossbands. The female is bright grass-green. The male has white vertebral spots, which are usually absent in the female. Both sexes have a white, whitish-blue, or whitish-yellow ventrolateral stripe; this is occasionally red in the male. The postocular streak is whitish-yellow and faint in the male and absent in the female; the absence of red coloration in this streak of both sexes separates the species from many conspecifics. It has a short spinose hemipenis. **Behaviour:** Nocturnal and arboreal. Often found near water, and often within a few metres of the ground. Feeds primarily on frogs, though reptiles and insects are taken. **Habitat:** Inhabits dry evergreen forest from 200 m to 1,200 m. **Range:** Occurs in northeastern (southernmost) and southeastern Thailand, being recorded in Chanthaburi, Nakhon Ratchasima, Prachinburi, and Trat provinces. **Regional Range:** Laos, Cambodia, and Thailand.

342

Pit Vipers Genus *Parias*

Members of this group are small to medium-sized snakes, all of which have heads very distinct from the neck. The eye has a vertical pupil. The chinshields are smooth. The tail may be short or moderately long and prehensile. They are crepuscular or nocturnal, with arboreal habits. Most species are ovoviviparous. Two species occur in Thailand.

343 Sumatran Pit-Viper *Parias sumatranus*

Body Length: Up to 256 cm.

Green pit viper with striking clean white lines along the entire length of the lower sides. Terminal part of tail reddish. **Behaviour:** Nocturnal and arboreal. Diet consists of rodents, birds, lizards, and frogs. **Habitat:** Occurs in Malayan mixed dipterocarp forest. **Range:** Found in extreme southern Thailand. Recorded in Pattani and Narathiwat provinces. **Regional Range:** Thailand, Peninsular Malaysia, and Indonesia (Sumatra).

343

344 Hagen's Green Pit-Viper *Parias hageni*

Body Length: Up to 96.0 cm.

Green pit viper with striking clean white lines along entire length of lower sides. There is a pink streak behind the eye. Chin and gular regions are pale green. Venter is pale green. **Behaviour:** Nocturnal and arboreal. Diet consists of rodents and birds. **Habitat:** Found in peninsular monsoonal evergreen forest and Malayan mixed dipterocarp forest. **Range:** Found in peninsular Thailand. Recorded in Surat Thani, Trang, and Songkhla provinces. **Regional Range:** Thailand, Peninsular Malaysia, and Indonesia (Sumatra).

344

Green Pit Vipers Genus *Popeia*

Members of this group are small to medium-sized snakes, all of which have very distinct heads from the neck. The eye has a vertical pupil. The chinshields are smooth. The tail may be short or moderately long and prehensile.

The sub-caudals are usually divided. They are crepuscular or nocturnal, with arboreal habits. Most species are ovoviviparous. Three species occur in Thailand.

345 Pope's Green Pit Viper *Popeia popeiorum*

Body Length: Up to 83.6 cm.

Uniformly bright green–coloured pit viper, lacking darker crossbands on the scales. The male has a conspicuous bi-colour streak behind the eye, thin and white below, wide and bright-red above; in the female the streak is white and thin in some specimens, and entirely missing in others. Males have a vivid bi-coloured, ventrolateral stripe, bright and deep red below, white above; females have only a well-defined white stripe. The tail is brown, with mottled green. Eyes are fire-red in both sexes. **Behaviour:** Nocturnal and arboreal. Diet consists of small mammals, birds, lizards, and frogs. **Habitat:** Often found in low vegetation and dense undergrowth. Occurs in sub-montane and montane forests. **Range:** Found in northern and western Thailand. **Regional Range:** India, Bangladesh, Myanmar, and Thailand.

345

346 Banded Pit Viper *Popeia fucatus*

Body Length: Up to 62.0 cm.

Dorsal colour is deep grass-green, with numerous white vertebral spots regularly spaced throughout the body every 2–4 vertebral scales and faint rusty-red crossbars on the 5–7 upper dorsal scale-rows. A well-defined bi-coloured ventrolateral stripe. The anterior and lower parts of the first dorsal scale-row are coral-red, the posterior and upper parts of this row and the lower part of second dorsal scale-row are pure white, running from the neck to the first third of the tail. The tail is greyish-green, heavily mottled on its side and above with large, irregular, rusty-brown blotches, then reddish-brown at its posterior part with darker blothes. The head is like the body, paler yellowish-green on the supra-labials, a little bit darker on the temporals. A very faint and discontinuous white postocular streak runs along the limit between the first and second row of temporals from the rear-most upper part of sub-ocular backward up to the angle of the mouth. Two or three very faint reddish-brown dots are present above the white streak. Eyes are deep yellowish-copper. **Behaviour:** Nocturnal and arboreal. Feeds on small mammals and birds. **Habitat:** Found in peninsular monsoonal evergreen forest and Malayan mixed dipterocarp forest. **Range:** Found throughout peninsular Thailand. **Regional Range:** Myanmar, Thailand, and Peninsular Malaysia.

346

347 Clouded Pit Viper *Popeia nebularis*

Body Length: Up to 88.1 cm.

Background colour is uniformly emerald-green, grass-green with some blue hue, or blue-green. A ventrolateral stripe, white or pale blue in males, absent or white and faint in females, extends from the angle of the mouth through first third-to-half of the tail on the first dorsal scale-row. The interstitial skin is very dark grey. The tail surface is basically the same colour as the dorsum on its sides, dark reddish-brown above, with a clear border between the two areas. The dorsal head surface and temporal regions are uniformly green, paler bluish-green on the supralabials. No postocular streak. The chin, throat, and venter are green or greenish-yellow, paler than upper head surface or sometimes bright yellow. Eyes are green or slightly yellowish-green in both sexes and juveniles. **Behaviour:** Arboreal. Diet presumably consists of rodents, birds, and lizards. **Habitat:** Found in Malayan

347

mixed dipterocarp forest. **Range:** Found in extreme south, in Narathiwat Province. **Regional Range:** Thailand and Peninsular Malaysia.

Green Pit Vipers Genus *Trimeresurus*

Members of this group are small to medium-sized snakes, all of which have heads very distinct from the neck. The eye has a vertical pupil. The chin-shields are smooth. The tail may be short or moderately long and prehensile. The sub-caudals are usually divided. They are crepuscular or nocturnal, with arboreal habits. Most species are ovoviviparous. One species occurs in Thailand.

348 Wirot's Pit Viper *Trimeresurus wiroti*

Body Length: Up to 88.9 cm.

An unusual overall-dark or very dark-patterned pit viper, with 22-35 darker crossbands. The male has various shades of dark greyish-brown, with dark brown irregular dorsolateral blotches. In the female, the patterning is less complex, and is often as dark as if not darker than in males. Females have shades of dark brown ground colour with darker subrectangular dorsolateral blotches. Both sexes have a distinctly projected and raised snout. The hemipenis is short with spines. The ventrals number 159-167 in males and 158-167 in females. **Behaviour:** Nocturnal and arboreal. Feeds on small mammals, birds, and frogs. **Habitat:** Inhabits peninsular monsoonal evergreen forest, Malayan mixed dipterocarp forest and peat swamp forest. **Range:** Recorded in Ron Phibun, Phattalung Province, and Phru To Daeng, Narathiwat Province. **Regional Range:** Thailand and Peninsular Malaysia.

348

Green Pit Vipers Genus *Tropidolaemus*

Members of this group are characterized by the absence of nasal pores. They have small, strongly keeled scales on the head and keeled gular scales. They are crepuscular and nocturnal, living in bushes and shrubs. One species occurs in Thailand.

349 Wagler's Pit Viper *Tropidolaemus wagleri*

Body Length: Up to 32.6 cm.

Very large head is strongly triangular. Neck is extremely narrow, being less than a third of the width of the head. It has a yellow or yellowish-white line on each side of the head. The very dark green body bears 25 small paired yellow spots usually edged with reddish. The prehensile tail is green with paired yellow or reddish spots **Behaviour:** Nocturnal and arboreal, living in understorey vegetation. Preys on small mammals, birds, lizards, and frogs. **Habitat:** Found in peninsular monsoonal evergreen forest, Malayan mixed dipterocarp forest, mangrove, and peat swamp forest. **Range:** Found in peninsular Thailand south of the Isthmus of Kra. **Regional Range:** Thailand, Peninsular and Eastern Malaysia, Singapore, Brunei, Indonesia, and Philippines.

349

Crocodilians

Order *Crocodylia*

Contains three families, namely the alligators and caimans (Alligatoridae), the crocodiles (Crocodylidae), and the gharial (Gavialidae). Worldwide: 23 species.

Crocodiles: Family Crocodylidae

These crocodilians are unspecialized with a moderately tapered snout that is not sharply distinct from the neck. The lower teeth generally fit into pits in the upper jaw, but the 4th tooth fits into a notch in the upper jaw, and is visible when the mouth is closed. The ventral scales have sensory pits. Worldwide: 14 species. Thailand: 3 species.

KEY TO THE THAI CROCODYLIDAE

1 a	Snout 3.0 to 3.5 as long as broad at the base; nuchal and dorsal scutes forming a single continuous shield composed of 22 transverse series, the broadest of which contains 6 scutes, the 4 anterior (nuchals) only 2 scutes—all keeled	*Tomistoma schlegelii*
1 b	Snout more than 1.5 and not more than 2.25 times as long as broad at the base	**2**
2 a	No longtitudinal ridge in front of eye; anterior nuchal scutes (postoccipitals) well-developed; a longitudinal ridge between orbits, none on snout	*Crocodylus siamensis*
b	A longitudinal ridge in front of eye; anterior nuchal series usually absent	*Crocodylus porosus*

Key based upon Chan-ard, Parr and Nabhitabhata

Crocodiles Genus *Crocodylus*

Members of this group have nine symmetrical head shields. Two species occur in Thailand.

350 Saltwater Crocodile *Crocodylus porosus*

Head to Tail: Up to 6-7 m.

The largest living crocodilian. It has a relatively large head and a long snout with lachrymal ridges from the eyes pointing toward the nostrils. Absence of enlarged neck scales behind the head. The dorsal body is dark olive or brown above. The abdomen is uniformly cream, whitish below. Melanistic forms are known. Young individuals are paler, spotted, and marked with 4-5 black bands above. The fingers are webbed at the base. **Behaviour:** The diet is varied, including mammals and birds as well as fish. Smaller individuals feed on snakes, lizards, mudskippers, and crustaceans. It accounts for most crocodile attacks on humans. **Habitat:** Found in coastal water habitats with brackish water and the tidal sections of rivers. Also found in freshwater sections of rivers, and frequents inland swamps and marshes. **Range:** Isolated records from Surat Thani, Trang and Narathiwat provinces. **Regional Range:** India, Sri Lanka, Bangladesh, Myanmar, China, Vietnam, Cambodia, Thailand, Peninsular and Eastern Malaysia, Singapore, Brunei, Indonesia and Philippines, Papua New Guinea and Australia, and Pacific Islands (eastward until Vanuatu).

350

351 Siamese Crocodile *Crocodylus siamensis*

Head to Tail: Up to 3-4 m.

Medium-sized crocodile less than 4 m in length. It has a broader snout than the similar-looking Saltwater Crocodile. Broad, heavy body. Mature animals have wide heads with bony ridges at the back of the skull. Dorsal is dark olive or brown above, whitish or yellowish below. Young individuals are paler, spotted, and marked with black above. **Behaviour:** A generalized feeder, including fish, snakes, and frogs. **Habitat:** Formerly found in most lowland rivers and streams. **Range:** Restricted to individuals in isolated pockets (Khao Ang Ru Nai Wildlife Sanctuary in Chachaengsao Province, Kaeng Krachan National Park in Petchaburi Province, and Pang Sida National Park in Prachinburi Province. **Regional Range:** Myanmar, Vietnam, Laos, Cambodia, Thailand, Malaysia, Singapore (extinct), and Indonesia. **International Conservation Status:** IUCN: Critical; CITES: Appendix I.

351

Gharials Genus *Tomistoma*

Members of this group have very long, narrow snouts. One species occurs in Thailand.

352 **Sunda Gharial** *Tomistoma schlegelii*

Head to Tail: Up to 5 m.

A large species with a distinctive long, slender snout. It has a dark body, with broad black bands on the tail, and dark blotches on the body. The juvenile has a similar body pattern to the adult. **Behaviour:** Feeds on fish and small invertebrates. **Habitat:** Restricted primarily to freshwater swamps,

352

rivers, and lakes; may occasionally use burrows. Slow-moving water and heavily vegetated habitats seem to be preferred. **Range:** Occurs in southernmost Thailand, in Narathiwat province; possibly extinct. **Regional Range:** Thailand, Peninsular and Eastern Malaysia, and Indonesia. **International Conservation Status:** IUCN: Endangered; CITES: Appendix I.

Selected Bibliography

Adler, K., H.M. Smith, S.H. Prince, P. David, and D. Chiszar. 2000. Russell's Viper: *Daboia russelii*, not *Daboia russellii*, due to classical Latin rules. *Hamadryad*. 25(2): 83–85.

Angsirichinda, W. 1997. The Agamidae in Thailand. Part of Special Subject of Biology. Unpublished report to the Department of Biology, Faculty of Science, Chulalongkorn University., Bangkok: i–vii + 1–107.

Aranyavalai, V. 2003. *Species diversity and habitat characteristics of butterfly lizards (Leiolepis spp.) in Thailand*. Ph.D. dissertation, Faculty of Science, Chulalongkorn University, Bangkok.

Bauer, A.M., and I. Das. 1998. New Species of *Cnemaspis* (Reptilia: Gekkonidae) from Southeastern Thailand. *Copeia* (2): 439–444.

Bauer, A. M., M. Sumontha, and O.S.G. Pauwels. 2003. Two new species of *Cyrtodactylus* (Reptilia: Squamata: Gekkonidae) from Thailand. *Zootaxa* 376: 1–18.

Bauer, A. M., M. Sumontha, W. Grossman, O.S.G. Pauwels, and G. Vogel. 2004. A new species of *Dixonius* (Squamata: Gekkonidae) from Kanchanaburi Province, western Thailand. *Current Herpetology* 23(1): 17–26.

Bauer, A.M., M. Sumontha, and O.S.G. Pauwels. 2008. A new red-eyed Gekko (Reptilia: Gekkonidae) from Kanchanaburi Province, Thailand. *Zootaxa* 1750: 32–42.

Bauer, A.M., K. Kunya, M. Sumontha, P. Niyomwan, N. Panitvong, O.S.G. Pauwels, L. Chanhome, and T. Kunya. 2009. *Cyrtodactylus erythrops* (Squamata: Gekkonidae), a new cave-dwelling gecko from Mae Hong Son Province, Thailand. *Zootaxa* 2124: 51–62.

Bennett, D. 1996. *Warane der Welt. Welt der Warane*. Edition Chimaira, Frankfurt am Main. (English edition: *Monitors of the World*, 1997, Edition Chimaira, Frankfurt am Main.

Boehme, W. 1981. A New Lygosomine Skink from Thailand (Reptilia: Scincidae). *Bulletin of the Museum of Civico Storia Naturale di Verona*. 8: 375–382.

Boulenger, G.A. 1885. *Catalague of Lizards in the British Museum (Natural History)*. Second edition. Volume III. Lacertidae, Gerrhosauridae, Scincidae, Anelytropidae, Dibamidae, Chameeleontidae, British Museum (Natural History), London.

Boulenger, G.A. 1890. *The Fauna of British India, Including Ceylon and Burma. Reptilia and Batrachia*. Taylor & Francis. London.

Boulenger, G.A. 1912. *A Vertebrate Fauna of the Malay Peninsula from the Isthmus of Kra to Singapore, including the Adjacent Islands. Reptilia and Batrachia*. Taylor & Francis, London: I–XIII.

Bourret, R. 1936. *Les Serpents del'Indochine. II. Catalogue systematique descriptif.* Henri Basuyau et Cie. Toulouse.

Bourret, R. 1941. *Les Tortues de l' Indochine*. Institut Oceanographie l' Indochine. Hanoi.

Brongersma, L.D. 1932. Some notes on the genus *Hemiphyllodactylus* Bleeker. *Zooloogische Mededeelingen* 14: 1–13.

Brown, R.M. 1999. New species of parachute gecko (Squamata: Gekkonidae: Genus *Ptychozoon*) from northeastern Thailand and central Vietnam. *Copeia* 1999 (4): 990–1001.

Bussarawit, S., A.P. Rasmussen, and M. Andersen. 1989. A preliminary study on sea snakes (Hydrophiidae) from Phuket Harbour, Phuket Island, Thailand. *Natural History Bulletin of the Siam Society* 37(2): 209–225.

Chan-ard, T., K. Thirakhupt, and P.P. van Dijk. 1996. Observations on *Manouria impressa* at Phu Luang Wildlife Sanctuary, Northeastern Thailand. *Chelonian Conservation and Biology* 2(1): 109–113.

Chan-ard, T., W. Grossmann, A. Gumprecht, and K. Schulz. 1999. *Amphibians and Reptiles of Peninsular Malaysia and Thailand: An Illustrated Checklist*. Bushmaster, Wurselen, Germany.

Chan-ard, T., Y. Chuaynkern, S. Thong-aree, and S. Phopinit. On the new discoveries and confirmation of amphibians and reptiles from Thailand. *Natural History Journal of Chulalongkorn University*. In press.

Cheke, A.S. 1973. New Locality for *Tylototriton verrucosus*. *Natural History Bulletin of the Siam Society* 24(3 & 4): 467.

Cheke, A.S. 1973. Snakes at Chiang Mai University. *Natural History Bulletin of the Siam Society* 24(3 & 4): 469–471.

Cheke, A.S. 1973. Two Lizard records for Chiang Mai Province. *Natural History Bulletin of the Siam Society* 24(3 & 4): 471–473.

Cochran, D.M. 1930. The herpetological collections made by Dr. Hugh M. Smith in Siam from 1923 to 1929. *Proceedings of the US National Museum* 77: 1–39.

Cochran, D.M. 1927. New reptiles and batrachians collected by Dr. Hugh M. Smith in Siam. *Proceedings of the Biological Society of Washington* 40: 179–192.

Cochran, D.M. 1941. A new species of lizard from Siam. *Copeia* 4: 238–239.

Cox, M.J. 1991. *The Snakes of Thailand and Their Husbandry*. Krieger Publishing Co., Malabar, Florida.

Cox, M.J., P.P Van Dijk, J. Nabhitabhata, and K. Thirakhupt. 1998. *A Photographic Guide to Snakes and Other Reptiles of Peninsular Malaysia, Singapore and Thailand*. New Holland Publishers UK Ltd. London.

Cox, M.J., P.P. Van Dijk, J. Nabhitabhata, and K. Thirakhupt.1998. *A Photographic Guide to Snakes and other Reptiles of Peninsular Malaysia, Singapore and Thailand*. New Holland Publishers (UK) Ltd., London.

Darevsky, I.S., and L.A. Kupriyanova. 1993. Two new all-female lizard species of the genus *Leiolepis* Cuvier, 1829 from Thailand and Vietnam. *Herpetozoa* 6(1/2): 3–20.

Das, I. 1997. A new species of *Cyrtodactylus* from the Nicobar Islands, India. *Journal of Herpetology* 31: 375–382.

David, P., and G. Vogel. 1996. The Snakes of Sumatra. An Annotated Checklist and Key with Natural History Notes. Edition Chimaira, Frankfurt am Main.

David, P., N. Vidal, and O.S.G. Pauwels. 2001. A morphological study of Stejneger's pitviper *Trimeresurus stejnegeri* (Serpentes, Viperidae, Crotalinae) with the description of a new species from Thailand. *Russian Journal of Herpetology* 8(3): 205–222.

David, P., G. Vogel, O.S.G. Pauwels, and N. Vidal. 2002. Description of a new species of the genus *Trimeresurus* from Thailand, related to *Trimeresurus stejnegeri* Schmidt, 1925 (Serpentes, Crotalidae). *Natural History Journal of Chulalongkorn University* 2(1): 5–19.

David, P., G. Vogel, and J. van Rooijen. 2008. A revision of the *Oligodon taeniatus* (Gunther, 1861) group (Squamata: Colubridae), with the description of three new species from the Indochinese Region. *Zootaxa* 1965: 1–49.

Dring, J.C.M. 1979. Amphibians and reptiles from Northern Trengganu, Malaysia, with descriptions of two new geckos: *Cnemaspis and Cyrtodactylus. Bulletin of British Museum of Natural. History (Zoology)* 34(5): 181–241.

Ellis, A.G. 1934. Python egg and young. *Natural History Bulletin of the Siam Society* 9(3): 329–330.

Ernst, C.H., and R.W. Barbour. 1989. Turtles of the World. Washington D.C.: Smithsonian Institution Press.

Flower, S.S. 1899. Notes on a second collection of reptiles made in the Malay Peninsula and Siam from November 1896 to September 1898, with a list of the species recorded from those countries. *Proceedings of the ZoologicalSociety of London* 1899: 600–679.

Frith, C.B. 1974. Second record of the sea snake *Laticauda colubrina* in Thailand waters. *Natural History Bulletin of the Siam Society* 25: 209.

Frith, C.B. 1975. A case of snake bite by the shore Pit Viper, *Trimeresurus purpureomaculatus* (Viperidae). *Natural History Bulletin of the Siam Society* 26: 159–169.

Frith, C.B. 1977a. A survey of the snakes of Phuket Island and the adjacent mainland areas of peninsular Thailand. *Natural History Bulletin of the Siam Society* 26: 263–316.

Frith, C.B. 1977b. The sea snake *Hydrophis spiralis* (Shaw): a new species of the fauna of Thailand. *Natural History Bulletin of the Siam Society* 26: 339–344.

Frith, C.B. 1978. Additions to the snake fauna of Phuket Island, peninsular Thailand. *Natural History Bulletin of the Siam Society* 27: 181–186.

Frith, C.B. 1987. Persian Gulf Sea Snake Hydrophis lapemoides (GRAY): new records from Phuket Island, Andaman Sea, and the Southern Part of the Straits Malacca. *Natural History Bulletin of the Siam Society* 35: 57–58.

Fritz, U., and F.J. Obst. 1999. Neue Schildkroten aus Sudostasien Teil II. Bataguridae (Cyclemys, Heosemys, Mauremys, Ocadia, Pyxidea, Sacalia) und Trionychidae. *Sauria* 21(1): 11–26.

Fritz, U., Guicking, D., Auer, M., Sommer, R. S., Wink, M., and A. K. Hundsdörfer. 2008. Diversity of the Southeast Asian leaf turtle genus *Cyclemys*: how many leaves on its tree of life? *Zoologica Scripta* 37: 367–390.

Gairder, K.G., and M.A. Smith. 1915. List of the mammals, birds, reptiles and bratrachians obtained in the Ratburi and Petchaburi Districts. *Natural History Bulletin of the Siam Society* 1(3): 146–156.

Golay, P., H.M. Smith, D.G. Broadley, J.R. Dixon, C. McCarthy, J.-C. Rage, B. Schatti, and M. Toriba. 1993. *Endoplyphs and other major venomous snakes of the world.* A checklist. Azemiops S. A. Herpetological Data Center, Aine-Geneve. 1–478.

Greer, A. 1997. *Leptoseps*: A New Genus of Scincid Lizards from Southeast Asia. *Journal of Herpetology* 31(3): 393–398.

Grismer, L.L., K.O. Chan, N. Nasir, and M. Sumontha. 2008. A new species of karst dwelling gecko (genus Strauch 1887) from the border region of Thailand and Peninsular Malaysia. *Zootaxa* 1875: 51–68.

Grismer, L.L., M. Sumontha, M. Cota, J.L. Grismer, P.L. Wood Jr., O.S.G. Pauwels, and K. Kunya. 2010. A revision and redescription of the rock gecko *Cnemaspis siamensis* (Smith, 1925) (Squamata: Gekkonidae) from Peninsular Thailand with description of seven new species. *Zootaxa* 2576: 1–55.

Grossmann W., and F. Tillack. 2000. Bemerkungen zur Herpetofauna des Khao Lak, Phang Nga, thailandische Halbinsel. Teil I: Einfuhrung; Amphibia; Reptilia: Sauria. *Sauria* 22(4): 23–38.

Grossmann W., and F. Tillack. 2001. Bemerkungen zur Herpetofauna des Khao Lak, Phang Nga, thailandische Halbinsel. Teil III: Ergebnisse der Jahre 1999 und 2000. *Sauria* 23(3): 21–34.

Grossmann, W., and T. Ulber. 1990. Ein neuer Gecko aus Zentral-Thailand: Gekko siamensis sp. Nov. (Reptilia: Sauria: Gekkonidae). *Sauria*. 12(3): 9–18.

Gumprecht, A. 1997. Die Bambusottern der Gattung *Trimeresurus* LACE'PEDE. Teil I: Die Chinesische Bambusotter *Stejnegeri* SCHMIDT, 1925. *Sauria*. 19(3): 9–30.

Gumprecht, A. 1998. Die Bambusottern der Gattung *Trimeresurus* LACE'PEDE. Teil II: Die Großaugen-Bambusotter *Trimeresurus macrops* KRAMER, 1977. *Sauria*. 20(3): 25–36.

Gumprecht A., and K. Tepedeleu. 1999. Die Bambusottern der Gattung *Trimeresurus* LACEPEDE Teil III: Die Java-Palmenotter *Trimeresurus puniceus* (BOIE, 1827). *Sauria*. 21(4): 25–36.

Gunther, A. 1860. On the reptiles of Siam. *Proceedings of the Zoological Society*. London. 113–117.

Heatwole, H. 1999. *Sea Snakes*. 2nd Edition. Krieger Publ. Co., Malabar, Florida.

Heyer, W.R. 1971. Description of some tadpoles from Thailand *Fieldiana Zoology*. 58(7): 83–91.

Hikida, T., N.L. Orlov, J. Nabhitabhata and H.Ota. 2002. Three new depressed-bodied water skinks of the genus *Tropidophorus* (Lacertilia: Scincidae) from Thailand and Vietnam. *Current Herpetology* 21(1): 9–23.

Honda, M., J. Nabhitabhata, H. Ota, and T. Hikida. 1997. A New Species of *Dibamus* (Squamata: Dibamidae) from Thailand. *Raffles Bulletin of Zoology* 45(2): 276–279.

Inger, R.F., and R.K. Colwell. 1977. Organization of contiguous communities of amphibians and reptiles in Thailand. *Ecological Monograph* 47(3): 229–253.

Inger, R.F. 1983. Morphological and ecological variation in the flying lizards (genus *Draco*). *Fieldiana Zoology*. new ser., no. 18. 1-35.

Iskandar, D.T. 2000. *Turtles and Crocodiles of Insular Southeast Asia and New Guinea*. PALMedia Citra, Bandung, Indonesia.

Iverson, J.B., and W.P. McCord. 1997. A new species of *Cyclemys* (Testudines: Bataguridae) from Southeast Asia. *Proceedings of the Biological Society of Washington* 110(4): 629–639.

Iverson, J.B., P.Q. Spinks., H.B. Shaffer, W.P. McCord, and I. Das. 2001. Phylogenetic relationships among the Asian tortoises of the genus Indotestudo (Reptilia: Chelonii: Testudinidae). Hamadryad 26(2): 283–287.

Karns, D.R., H.K. Voris, T. Chan-ard, J.C. Goodwin, and J.C. Murphy. 1999-2000. The spatial ecology of the Rainbow Water Snake, *Enhydris enhydris* (Homalopsis) in Southern Thailand. *Herpetological Natural History* 7(2): 97–115.

Kharin, V.E. 1984. A review of the sea snakes of the group *Hydrophis* sensu lato (Serpentes: Hydrophiidae) 3. The genus *Leioselasma*. *Zoological Zhurnal*. 63: 1536-46 (in Russian).

Kharin, V.E., and Cheblukov. V.P. 2007. On First Reliable Record of the Sea Snake *Chitulia belcheri* (Gray. 1849) from Australian Waters, with Notes on Species Composition and Taxonomic Status of the Genus *Chitulia* (Serpentes, Hydrophiidae). *Russian Journal of.Marine Biology*. Vol.33: 161–165.

Kramer, E. 1977. Zur Schlangenfauna Nepals. *Revue Suisse de Zoologie* 84(3): 721–761.

Kroon, C. 1973. A new Colubrid Snake (Boiga) from Southeastern Asia. *Copeia* 1973(3): 580–586.

Lang, M., and W. Böhme. 1990. Description and phylogenetic position of a new species of *Isopachys* from central Thailand and Northern Burma (Squamata: Scincidae). *Bulletin de l'Institute Royal des Sciences Naturelles de Belgique Biologie* 60: 231–240.

Lauprasert, K., and K. Thirakhupt. 2001. Species diversity, distribution and proposed status of Moniter Lizard (Family Varanidae) in Southern Thailand. *Natural History Journal of Chulalongkorn University* 1(1): 39–46.

Lonnberg, E. 1916. Zoological results of the Swedish Zoological Expedition to Siam. 2. Lizards. *Kongl. Svenska Vetenskaps Academiens Handlingar* 55(4): 1–12.

Manthey, U., and W. Grossmann. 1997. *Amphibien und Reptilien Sudostasiens*. Natur und Tier-Verlag, Munster.

Manthey, U., and J. Nabhitabhata. 1991. Eine new Agame, *Ptyctolaemus phuwuanensis* sp. n. (Sauria: Agamidae), aus Nordest-Thailand. *Sauria*. 13(4): 3–6.

Marquez, M. 1990. *Sea Turtles of the World*. An annotated and illustrated catalogue of sea turtle species known to date. *FAO species catalogue*. Vol.11. Rome.

McCord, W.P., and P.C.H. Pritchard. 2002. A review of the softshell turtles of the Genus *Chitra*, with the description of new taxa from Myanmar and Indonesia (Java). *Hamadryad* 27(1):11–56.

Murphy, J.C. 2007. *Homalopsid Snakes: Evolution in the Mud*. Krieger Publishing Company, Malabar, Florida.

Murphy, J.C., H. Voris, D.R. Karns, T. Chan-ard, and K. Suvunrat. 1999. The ecology of the water snakes of Ban Tha Hin, Songkhla Province, Thailand. *Natural History Bulletin of the Siam Society* 47: 129–149.

Murphy, J.C., T. Chan-ard, S. Makchai, M. Cota, and H. Voris. 2008. The rediscovery of Angel's stream snake, *Paratapinophis praemaxillaris* Angel, 1929. (Reptilia: Serpentes: Natricidae). *Natural History Journal of Chulalongkorn University* 8(2): 160–183.

Musters, C.J.M. 1983. Taxonomy of the Genus *Draco* L. (Agamidae, Lacertilia, Reptilia). *Zoologische Verhandelingen*. Leiden (199): 1–120.

Nabhitabhata, J., T. Chan-ard, and Y. Chuaynkern. 2000. Checklist of Amphibians and Reptiles in Thailand. Office of Environmental Policy and Planning, Bangkok.

Niyomwan, P. 1999. *Species Diversity, Morphology and Habitat Types of Blind Snakes (Family Typhlopidae) in Thailand*. Chulalongkorn University, Bangkok.

Nutaphand, W. 1979. *The Turtles of Thailand*. Bangkok. Siamfarm Zoological Garden.

Ota, H., T. Hikida, J. Nabhitabhata, and S. Panha. 2001. Cryptic taxonomic diversity in two broadly distributed lizards of Thailand (*Mabuya macularia* and *Dixonius siamensis*) as revealed by chromosomal investigations (Reptilia: Lacertilia). *Natural History Journal of Chulalongkorn University* 1(1): 1–7.

Ota, H. and J. Nabhitabhata. 1991. A new species of *Gekko* (Gekkonidae: Squamata) from Thailand. *Copeia* (2): 503–509.

Ouboter, P.E. 1986. A Revision of the Genus *Scincella* (Reptilia: Sauria: Scincidae) of Asia, with some notes on its Evolution. *Zoologische Verhandelingen*. 229: 1–100.

Pauwels, O.S.G., Wallach, V., David, P., and L. Chanhome. 2002. A New Species of *Oligodon* Fitzinger, 1826 (Serpentes, Colubridae) from Southern Peninsular Thailand. *Natural History Journal of Chulalongkorn University* 2(2): 7–18.

Pauwels, O.S.G., O.-A. Laohawat, W. Naaktae, C. Puangjit, T. Wisutharom, C. Chimsunchart, and P. David. 2002. Reptile and amphibian diversity in Phang-nga Province, southern Thailand. *Natural History Journal of Chulalongkorn University* 2(1): 25-30.

Pauwels, O.S.G., M. Sumontha, P. David, T. Chan-ard, and G. Vogel. 2005. First Record of *Lycodon cadamomensis* from Thailand, with a key to Thai *Lycodon* (Serpentes: Colubridae). *Natural History Bulletin of the Siam Society* 53(1): 79–86.

Penyapol, A. 1958. A preliminary Study of the Sea Turtles in the Gulf of Thailand. *Natural History Bulletin of the Siam Society* 17: 23–37.

Peters, Van G. 1971. Die intragenerischen Gruppen und die Phylogenese der Schmetterlingsagamen (Agamidae: *Leiolepis*). *Zoological Journal of Systematics*. 98: 11–152.

Popinit, S. 1991. Species Diversity and Distribution of Reptiles in Suborder Lacertilia at Phu Khieo Wildlife Sanctuary Changwat Chaiyaphum. Ph.D. dissertation Kasetsart University.

Purananda, C. 1957. Poisonous snakes in Siam. *Natural History Bulletin of the Siam Society* 18: 1–14.

Rao, D.K., and D.T. Yang. 1992. Phylogenetic systematic of Pareatinae (Serpents) of Southeastern Asia and adjacent islands with relationship between it and the geology change. (Chinese, with English summary) *Acta Zoologica Sinica* 38(2): 39–150.

Rasmussen, A.R. 1987. Persian Gulf sea snakes *Hydrophis lapemoides* (Gray): new record from Phuket Island, Andaman Sea, and the southern part of the Straits of Malacca. *Natural History Bulletin of the Siam Society* 35(1): 57–58.

Rasmussen, A.R. 1989. Sea Snakes *Thalassophina viperina* (Schmidt) and *Laticauda laticaudata* (Linn.): New records from Phuket Island, Andaman Sea, with remarks on the two subspecies of *L. laticaudata*. *Natural History Bulletin of the Siam Society* 37(1): 99–103.

Rasmussen, A.R. 2001. Sea Snakes. In: Carpenter, K.E. and Niem, V.H. (Eds) *FAO species identification guide for fishery purposes. The living marine resources of the Western Central Pacific: Volume 6*. Food and Agriculture Organization of the United Nations, Rome.

Robinson, H.C., and C.B. Kloss. 1914. Reptiles and batrachians from Bandon, Koh Samui and Koh Pennen. *Journal ofFederal Malay States Museum* 1(3): 153–155.

Round, P.D., and U. Treesucon. 1989. A record of the Red Headed Krait, *Bungarus flaviceps*, in Ranong Province, Southern Thailand. *Natural History Bulletin of the Siam Society* 37(2): 225.

Saint Girons, S.H. 1972. *Les Serpents du Cambodge*. Memoires du Museum National D' Histoire Naturelle. Serie A, Tome LXXIV: 1–170.

Schafer, C. 1992. *Gekko monarchus*. *Sauria* 14(1–4): 243–246.

Schulz, K.-D. 1996. *A Monograph of the Colubrid Snakes of the Genus* Elaphe *Fitzinger*. Koeltz Scientific Books, Havlickuv Brod.

Siah, Y. 1957. Siamese crocodile. *Natural History Bulletin of the Siam Society* 18: 31–36.

Slowinski, J.B., J. Boundy, and R. Lawson. 2001. The phylogenetic relationships of Asian coral snakes (Elapidae: *Calliophis* and *Maticora*) based on morphological and molecular characters. *Herpetologica* 57(2): 233–245.

Slowinski, J.B., and W. Wuster. 2001. A new cobra (Elapidae: *Naja*) from Myanmar (Burma). *Herpetologica* 56(2): 257–270.

Smith, M.A. 1912. A list of the crocodiles, tortoise, turtles, and lizards at present known to inhabit Siam. *Natural History Bulletin of the Siam Society* 21: 48–57.

Smith, M.A. 1914. The snakes of Bangkok. *Natural History Bulletin of the Siam Society* 1: 5–18.

Smith, M.A. 1914. The snakes of Bangkok. *Natural History Bulletin of the Siam Society* 2: 93–104.

Smith, M.A. 1914. The distribution of *Ancistrodon rhodostoma* the Malayan viper in Siam. *Natural History Bulletin of the Siam Society* 1: 57–58.

Smith, M.A. 1914. Large banded krait. *Natural History Bulletin of the Siam Society* 1(1): 58–59.

Smith, M.A. 1914. On the breeding habits of *Hypsirhina enhydris* and *Herpeton tentaculatum*, the tentacle snake. *Natural History Bulletin of the Siam Society* 1(1): 126–127.

Smith, M.A. 1915. The snakes of Bangkok. Part III, with 4 plates. *Natural History Bulletin of the Siam Society*1(3): 173–187.
Smith, M.A. 1915. On the breeding and colour changes in the lizard, *Calotes mystaceus*. *Natural History Bulletin of the Siam Society* 1(4): 256–257.
Smith, M.A. 1916. On a collection of reptiles and batrachians from peninsular Siam. *Natural History Bulletin of the Siam Society* 2(2): 148–171.
Smith, M.A. 1916. Descriptions of three new lizards and a new snake from Siam. *Natural History Bulletin of the Siam Society* 2(1): 44–47.
Smith, M.A. 1917. Description of new reptiles and a new batrachian from Siam. *Natural History Bulletin of the Siam Society* 2(3): 221–225.
Smith, M.A. 1917. Description of a new snake and a new frog from Siam. *Natural History Bulletin of the Siam Society* 2(4): 276–278.
Smith, M.A. 1918. Description of new snake *Opisthotropis spenceri* from Siam. *Natural History Bulletin of the Siam Society* 3(1): 13.
Smith, M.A. 1919. *Crocodilus siamensis. Natural History Bulletin of the Siam Society* 3(3): 217–221.
Smith, M.A. 1919. The lizards of the genus *Tropidophorus* in Siam with descriptions of two new species. *Natural History Bulletin of the Siam Society* 3(3): 223–228.
Smith, M.A. 1921. Reptiles and batrachians collected on Pulo Condore. *Natural History Bulletin of the Siam Society* 4(2): 93–97.
Smith, M.A. 1923. The poisonous land snakes of Siam. *Natural History Bulletin of the Siam Society* 6(1): 55–64.
Smith, M.A. 1923. Note on reptiles and batrachians from Siam and Indo-China. (No. 2). *Natural History Bulletin of the Siam Society* 6 (1): 47–53.
Smith, M.A. 1926. *Monograph of the Sea Snakes (Hydrophiidae)*. British Museum, London.
Smith, M.A. 1929. Remarks on three rare reptiles from the Indo-chinese Region. *Natural History Bulletin of the Siam Society* 8(1): 49–50.
Smith, M.A. 1930. Herpetological collections made in Siam from 1923 to 1929. *Natural History Bulletin of the Siam Society* 8(2): 126–127.
Smith, M.A. 1931. *The Fauna of British India, including Ceylon and Burma: Reptilia and Amphibia*. Vol 1. *Loricata, Testudines*. Taylor & Francis Ltd, London.
Smith, M.A. 1935. *The Fauna of British India, including Ceylon and Burma*. Reptilia and Amphibia, Vol. 2. *Sauria*. Taylor & Francis Ltd, London.
Smith, M.A. 1943. *The Fauna of British India, Ceylon and Burma, Including the Whole of the Indo-Chinese Subregion. Reptilia and Amphibia*. Vol.3. *Serpentes*. Taylor & Francis Ltd. London.
Smith, M.A., and B.K. Kloss. 1915. On reptiles and batrachians from the coast and islands of south-east Siam. *Natural History Bulletin of the Siam Society* 1(3): 237–249.
Soderberg, P. 1966. Two little known snakes from Khao Yai. *Natural History Bulletin of the Siam Society* 21: 34–40.
Soderberg, P. 1967. The newt *Trilototritron verrucosus* Anderson, 1871, in Thailand. *Natural History Bulletin of the Siam Society* 22: 211–212.
Taylor, E.H. 1920. Philippine Turtles. *Philippine Journal of Science* 15: 1–144.
Taylor, E.H. 1934. Zoological results of the Third de Schanensee Siamese Expedition, pt. III. Amphibians and Reptiles. *Proceedings of Academic National Science Philadelphia*. 86: 281–310.
Taylor, E.H. 1962. New Oriental Reptiles. *University of Kansas Science Bulletin* 43(7): 209–263.
Taylor, E.H. 1963. The lizards of Thailand. *University of Kansas Science Bulletin* 44(14): 687–1077.

Taylor, E.H. 1965. The serpents of Thailand and adjacent water. *University of Kansas ScienceBulletin* 45(9): 609–1096.

Taylor, E.H. 1970. The turtles and crocodiles of Thailand and adjacent water with a synoptic herpetological bibliography. *University of Kansas Science Bulletin.* 49(3): 89–179.

Taylor, E.H. 1971. The caecilian fauna of Thailand with an examination of scale characters. *Natural History Bulletin of the Siam Society* 24: 33–39.

Taylor, E. H., and R. E. Elbel. 1958. Contribution to the herpetology of Thailand. *Kansas University Science Bulletin.* 38: 1033–1189.

Tweedie, M.W.F. 1983. The Snakes of Malaya. Singapore National Printers (Pte) Ltd., Singapore.

Ulber, T. 1993. Bemerkungen über cyrtodactylone Geckos aus Thailand nebst Beschreibungen von zwei neuen Arten (Reptilia: Gekkonidae). *Mitteilungen aus dem Zoologischen Museum in Berlin* 69(2): 187–200.

Ulber, T., and W. Grossman. 1991. Ein weiterer neuer Gecko aus Zentral-Thailand: *Cyrtodactylus papilionoides* (Reptilia: Sauria: Gekkonidae). *Sauria* 13(1): 11–20.

Wallach, V. 2001. *Typhlops roxaneae*, a new species of Thai blindsnake from the *T. diardii* species group, with a synopsis of the Typhlopidae of Thailand (Serpentes: Scolecophidia)., *Raffles Bulletin of Zoology* 49(1): 39–49.

Warrell, D. A., S. Looareesuwan, A. F. Stimson, and R. A. Hutton. 1992. Rediscovery and redefinition of Malcolm Smith's *Trimeresurus kanburiensis* in Thailand, with a report of envenoming. *Transactions of the Royal Society of Tropical Medicine and Hygiene* 86: 95–99.

Way, W.K. 1975. Re-discovery of *Riopa haroldyoungi* Taylor, 1962. *Natural History Bulletin of the Siam Society* 26: 159–169.

Welch, K.R.G. 1988. *Snakes of the Oriental. A Checklist.* Krieger Publ. Co., Malabar, Florida.

Welch, K.R.G., P.S. Cooke, and A.S. Wright, 1990. *Lizards of the Orient: A Checklist.* Malabar: Robert E. Kreiger Publishing Co.

Wongratana, T. 1984. Range extension of the crocodile salamander, *Tylototriton verrucosus*, to Phu Luang, Thailand. *Natural History Bulletin of the Siam Society* 32(2): 107–110.

Wuster, W., and R.S. Thorpe. 1990. Systematics and biogeography of the Asiatic Cobra (*Naja naja*) species complex in the Philippine Islands. In G. Peters and R. Hutterer (Eds.)Vertebrates in the Tropics. Pp 333–344. Museum A. Koenig, Bonn.

Wuster, W., R.S. Thorpe, M.J. Cox, P. Jintakune, and J. Nabhitabhata. 1995. Population systematics of the snake genus *Naja* (Reotilia: Serpentes: Elapidae) in Indochina: multivariate morphometrics and comparative mitochondrial DNA sequenceing (cytochrome oxidase I). *Journal of Evolutionary Biology* 8: 493–510.

Yamakushi, T., T. Hikida, J. Nabhitabhata, S. Panha, and H. Ota. 2001. Geographic variations in the Common Skink *Sphenomorphus maculatus* (Blyth, 1853) in Thailand, with re-validation of *Lygosoma mitenense* Annandale, 1905 on its subspecies. *Natural History Journal of Chulalongkorn University* 1(1): 23–31.

Index of Scientific Names

The scientific names used in this book are listed below. The numbers refer to the species numbers assigned in this book.

Index of Common Names

The common names used in this book are listed below. The numbers refer to the species numbers assigned in this book.